The Parent's Guide to SCIENCE FAIRS

JOHN BARRON

ROXBURY PARK

LOWELL HOUSE

LOS ANGELES

NTC/Contemporary Publishing Group

Library of Congress Cataloging-in-Publication Data

Barron, John.
    The parent's guide to science fairs / John Barron.
        p.   cm.
    Includes bibliographical references and index.
    ISBN 0-7373-0269-0 (alk. paper)
    1. Science projects. 2. Science exhibitions. I. Title.
Q182.3.B37 1999
507'.8—dc21   99-38907
                                                                  CIP

Published by Lowell House
A division of NTC/Contemporary Publishing Group, Inc.
4255 West Touhy Avenue, Lincolnwood (Chicago), Illinois 60646-1975 U.S.A.

Lowell House books can be purchased at special discounts when ordered in bulk for
premiums and special sales. Contact Department CS at the following address:
NTC/Contemporary Publishing Group
4255 West Touhy Avenue
Lincolnwood, IL 60646-1975
1-800-323-4900

Roxbury Park is a division of NTC/Contemporary Publishing Group, Inc.

Managing Director and Publisher: Jack Artenstein
Editor in Chief, Roxbury Park Books: Michael Artenstein
Director of Publishing Services: Rena Copperman
Editorial Assistant: Nicole Monastirsky
Freelance Editor: Asha Johnson
Interior Designer: Carolyn Wendt

Printed and bound in the United States of America
99 00 01 DHD 10 9 8 7 6 5 4 3 2 1

To my daughter Lauren who in a few years will start on her own science fair journey.

CONTENTS

CHAPTER 1

What Is a Science Fair?

*A science fair is "an occasion for the display
and evaluation of student research projects."*
(The Ohio Academy of Science 1987, p. 3)

Picture this: your child has just come home from school, all excited, and says, "We're doing a science fair in school and I'm going to enter a project!" Welcome to the wonderful world of science fairs. You've just embarked on what may be a journey of eight or nine years: your child's science fair experience. The science fair is a chance for your son or daughter to actually do science, get hands-on experience, and learn by doing. It can get your child excited about school and also develop skills that will be useful for the rest of his or her academic and working careers. In this one activity, your child can meet almost half of the general objectives in the science course—and have fun doing it.

The power to create a positive experience for your child in this endeavor is within your hands, but don't worry, you're not alone. We're here to help.

The Ohio Academy of Science says it all in our opening quote, "an occasion for the display and evaluation of student research projects." Sounds kind of ominous, doesn't it? It doesn't have to be that way, even though a science fair is not just another day at school for your child. Picture a gymnasium or other large hall. Now imagine it filled with lines of tables like at a flea market or a bingo parlor. Finally, throw in a bunch of students with all their research wares proudly displayed. That's a science fair.

A science fair is a chance for students to do real scientific research and present it to their peers, their teachers, and the panel of expert judges. They get to show off—to many different people—all the good, hard work they've done. A science fair may be a small event encompassing just one science class in a school, or it may be large, such as the 600-participant Canada Wide Science Fair. In either case it's important to your child, to your child's education, and, we hope, to you.

WHY SHOULD PARENTS PARTICIPATE IN THEIR CHILD'S SCIENCE FAIR?

Why become involved in your child's science fair project, you may ask? The obvious reason is that you can affect how well your child does through your interest and help. Many educators feel that parental involvement can actually determine a child's success in a science fair. Two supporters of parents in the science fair process are Charles Pryor and Ava Pugh (1987). They suggest inviting parents to help out in the classroom during the project process. They involve parents heavily in the science fair project process with one important proviso: they have meetings with the parents to insure that they know just how far into the project they should go. Parents who get *too* involved are always a concern for science fair organizers. If at the end of the project you could put your own name on it, then you have done too much. Participate, but make sure you aren't interfering.

A science fair can become a family affair. Linda Sittig (1985) relates her family's experience with a science fair. The whole family had a meeting to determine the best way to undertake a given project idea while refining the idea itself. But she firmly believes that the idea had to come from her children and they had to do most of the work as well. Teachers sometimes receive letters from parents explaining how much help they gave their children and reassuring the teacher that their child did most of the project. Teacher Stephen Henderson (1983) received one such letter that told of a student who worked on his project with his grandfather, learning not only science but also important family history and personal skills. All in all, involving yourself in your child's science fair project is a good thing.

As competitive fairs appear at lower and lower grades and school curricula emphasize a more practical approach to science, the science fair is regaining the preeminence it enjoyed in the 1950s and 1960s. What this means from a parental point of view is that you can expect your child to be involved in science fairs in the school from as early as grade four to the end of her high school years. Only athletics and music study can command as wide a spectrum of your child's educational years as the science fair does. So when you embark on your child's first science fair experience, keep in mind that this is only the first step on a long and winding road full of bumps and potholes, but leading to learning and skills for your child.

There is much to a science fair project that the uninitiated would not understand. What is the scientific method? Why can't my child do this project? What is a research guideline? How can I help my child come up with an idea? Where do we go from here? These are just some of the questions that parents all over North America ask during science fair season. The answers to these questions and a whole lot more can be found within these pages. Enjoy!

Rules of the Fair

*The Intel ISEF is the Olympics, the World Series,
and the World Cup of science competition.*
(Science Service 1998)

Science fairs within the United States follow many and varied structures. Local school fairs may send their winners on to larger city-
wide fairs or to regional or statewide fairs. In Canada, the structure is
a bit more defined. Local fairs (usually at the school level) feed into
regional fairs (geographic regions, sometimes whole provinces), of
which there are 108. The best from these go on to the Canada Wide
Science Fair.

INTERNATIONAL SCIENCE AND ENGINEERING FAIR

In either country, winners may be sent on to the International Science and Engineering Fair (ISEF), the largest international science fair in the world. To send winners to this prestigious event, the fair must be "affiliated" with the ISEF. Affiliation basically means membership in the exclusive club of science fairs. To obtain membership or affiliation, a fair needs to consist of at least five participating high schools and/or fifty students in the ninth to twelfth grades (Science Service 1998). Currently over five hundred of these fairs exist all over the world. A complete list of recent affiliations is found in Appendix A.

Affiliation grants with it a privilege as well as a responsibility. The privilege is the ability to send the top projects from your fair to this international competition. The responsibility is to adhere to the reasonable but strict guidelines of the ISEF.

Unlike other fields that have governing bodies, such as sports associations, the science fair rules and structure face a more trickle-down approach. The ISEF is not a governing body over science fairs in the U.S. or Canada, but its rules have an effect in both countries due to local, regional, state, and national fairs being affiliated with it. In order for a fair to send its winners to the International Science and Engineering Fair, their projects must conform to ISEF's regulations.

Even if your fair is not affiliated with the ISEF, following their regulations—which are balanced, scientifically proper, and ethically correct—is not a bad idea. Some nonaffiliated fairs already do this, such as the Keithville Science Fair in Shreveport, Louisiana, which judges its fairs within ISEF guidelines (Silver 1994). Of course, for these rules to be this good they also need to be very detailed. They cover a significant amount of material, everything from proper scientific procedures with animals to exact project dimensions, so we will encapsulate them as best we can and simplify them as well.

DIVISIONS

The International Science and Engineering Fair has divided science fair competition into fifteen divisions. Each of these divisions corresponds directly to a real life subdivision of science and technology. This gives the fair a more real-world feel and makes it easier to assign judging teams as well. At a science fair, projects of the same division or type would compete against each other. At some fairs, projects of

---

The Fifteen Divisions

Behavioral and Social Sciences: dealing with human and animal behavior

Biochemistry: dealing with chemical life processes

Botany: dealing with plant life

Chemistry: dealing with matter composition

Computer Science: hardware or software development

Earth and Space Sciences: dealing with geological processes and the study of space

Engineering: practical application of science and technology

Environmental Science: the study of ecology and pollution

Gerontology: the study of aging

Mathematics: dealing with formal mathematical systems and their application

Medicine and Health: the study of human and animal health

Microbiology: dealing with the biology of microorganisms

Physics: the study of energy and its effects

Team Projects: any project involving more than one and fewer than four students

Zoology: the specific study of animals.

Source: Science Service 1998

---

the same grade level would compete against each other, but at the ISEF no distinction is made between grade levels—everyone competes as one great big science family.

Of course whatever idea your child chooses for his project will determine in which category he will compete. Team projects compete in their own separate category, which means that having a team project does not give the students an unfair advantage over single-student projects.

SCIENCE VERSUS TECHNOLOGY

In recent years, science fair projects have also involved engineering and computer-based projects. Without getting into a long, drawn-out discussion of the difference between science and technology, we will simply suppose that science is mostly theory, and technology is mostly the application of that theory. So if technology is part of the process, why don't we call it a science and technology fair? Well, many science fair organizers have rebelled against the idea, but the reality is that computer programs, graphics, and engineering displays will continue to have a role. The International Science and Engineering Fair recognizes the two main streams of its fair distinctly within its name.

DISPLAY REGULATIONS

At any science fair, there are certain things you can and cannot show, demonstrate, or do. These rules are set down for fairness and for safety within the fair site itself. Some items are unacceptable for display due to the obvious safety and health hazards associated with them. The chart below encapsulates the do's and don'ts for fair display. Please note that for your local fair, more or less stringent rules may apply based on local regulations and the space in which the fair is to be held. Check with your child's science teacher to make sure. Having an extensive project denied entry due to size or safety considerations on the morning of a fair is not a pleasant experience for the student—or the fair organizer.

CANNOT DISPLAY

living organisms	human/animal parts or fluids	sharp items
dried plant materials	lab chemicals and water	highly flammable items
preserved animals	hazardous chemicals or devices	topless batteries
foods		awards, flags, or business cards
soil or waste samples	sublimating solids	personal identifiers

CAN DISPLAY BUT NOT OPERATE

projects with unshielded moving parts	class III and class IV lasers	devices requiring voltage over 125 volts

Source: Science Service 1998

Along with safety considerations is a size limitation. A science fair is usually held in limited space—a school gym, an arena, or a stadium—and this space has to be allocated evenly for everyone. To ensure this, a size limitation for projects is imposed of 30 inches deep, 48 inches wide, and 108 inches high including the table it is resting on. Fear not, this space is quite sufficient for most backboard arrangements and project displays, but it is important to keep this limitation in the back of your mind at all times.

THE SCIENTIFIC REVIEW COMMITTEE

Just choosing a project idea and keeping within the safety and size limitations is not enough. As a parent you are the responsible individual in your child's project, and you have to ensure that ethical guidelines and rules are followed. Of course, the science teacher is another of these responsible individuals, and may undertake much of what we are about to discuss, but you should still be informed and know exactly what is going on.

At each level of your child's science fair experience, a scientific review committee will review the student's project to ensure that no rules—or heaven forbid, any laws—are broken. The scientific review committee is sometimes called different things at each level, a safety committee, an ethical research committee, a project review committee, whatever—they all do the same thing. Several people are responsible for ensuring that the student's project isn't tossed out by this committee due to some transgression. The student is the one who suffers most from the mistake and deserves some help in avoiding this. One of the roles of the teacher is to ensure that the project stays within ISEF guidelines throughout the project process (Silver 1994). But ultimately, the parents should have been following their child's research closely. As in the law, ignorance of the rules is no excuse. Fortunately, this book will help you avoid that pitfall.

Certain types of projects will require more than just teacher or parental supervision, they will require the involvement of a qualified scientist. There are always going to be situations where the student could use some expert help, but certain situations *require* expert help. Student experimentation involving human subjects, vertebrate animals (those with a backbone), pathogens (poisons, etc.), controlled substances (illegal drugs, weapons), recombinant DNA (the building blocks of all life), or human/animal tissue may require the help of a qualified scientist. In either case, such experimentation should be looked at closely by the scientific review committee (or whatever your fair calls it) before experimentation takes place. Federal and state agencies have strict laws and guidelines regarding experimentation involving some of these items, so be sure to check on this as well before beginning an experiment.

HUMAN AND ANIMAL RESEARCH

At the Canada Wide Science Fair, there is a saying about animal and human research, "If it hurts, you can't do it." Under this fair's regulations, no research can be undertaken that causes undue stress on a human or animal. The same is not true for the ISEF. The regulations governing human and animal experimentation allow undue stress to humans and animals as long as a qualified scientist supervises it. The reality of the scientific workplace is that this occurs thousands of times each day in the United States alone. The regulations also stipulate that although the research cannot lead to the death of an animal, it can lead to a point where the animal must be properly euthanized. Strict guidelines of animal care, such as those enforced at research labs, are required to be followed at all times during the life of the experiment. This housing and care cannot take place at home but rather at a recognized institution or school.

The ISEF rules, while not barring the use of vertebrate animals for experimentation, stress, however, that alternatives (such as cells, tissue cultures, plants, etc.) that would be equally valid for the research must be used instead of vertebrate animals. This points out the importance of a student checking with the scientific review committee, if only to avoid needless research.

PATHOGENIC AGENTS AND RECOMBINANT DNA

The ISEF allows students to undertake experiments using recombinant DNA (rDNA) and pathogenic agents as long as students remain within the bounds of federal law and guidelines. It is a good idea to know what your child is getting into before starting some of these experiments. Recombinant DNA (rDNA) is, in simple terms, a form of DNA with some strands removed and combined with strands from other DNA, hence the term recombinant or recombined. Of course, the ISEF requires that a scientific review committee must first approve this type of experimentation. On a local, regional, or state level this is also a good suggestion. Make sure that your fair organizers and safety committee are aware of this experiment taking place.

CONTROLLED SUBSTANCES

Experimentation with controlled substances (drugs, alcohol, and tobacco) must also adhere to the state and federal laws governing them. In some cases, this will mean that the planned experiment may not be viable right from the beginning. The student will also be required to work with a qualified scientist to do any sort of human or animal testing with these substances. Once again, scientific review committee approval is necessary for entering the ISEF, so this is something you should check out at the entry level of the project.

HAZARDOUS MATERIALS AND DEVICES

This section deals primarily with explosives, radioactive materials, and firearms. Hazardous chemicals also fall under this category: things such as pesticides, known carcinogens (cancer causing agents), toxic chemicals, and flammable chemicals. Experimentation with this type of material—even possession of it—will require permits and adherence to strict guidelines. In this area, the ISEF does not require a qualified scientist, but rather a "designated supervisor." The difference between the two is that the designated supervisor does not need an advanced degree but has to work with or be trained about the substance or material being used.

CONCLUSION

That covers most of the rules for a project entering the ISEF. Of course, each level of the science fair in your area will put its own spin on these rules and perhaps add more depending upon the situation, but if the fair is affiliated with the ISEF, following the rules outlined here will do your child no harm. Regardless, always check with your child's science teacher. If she can't answer your question, check with the organizers of your local, regional, or state fair. They are usually ready, willing, and able to answer any questions you may have.

How to Design a
Science Fair Project

A science fair project is a presentation of an experiment, a demonstration, a research effort, a collection of scientific items, or a display of scientific apparatus.
(Asimov and Fredericks 1990, p. 1)

Now that you know what you can and cannot do for a science fair project, it's time for the how-to section. The first step of any project, be it carpentry or science, is to establish a plan or process to follow. Any project will be at least presentable if you follow the basic plan of experimentation and research called the scientific method. Actually, there is no set "method" of science. Philosophers and scientists have been arguing this point since Aristotle's time. Everyone develops or discovers science in unique ways, so there can be no one "scientific method."

But we are going to share a plan that has been developed from years of experience, extensive reading, and a whole lot of trial and error (more of the latter than the former). It may not be the best possible plan, but it works. Many award-winning science fair projects have followed this plan.

THE PROCESS IN A NUTSHELL

1. *Idea generation:* come up with the idea for the project.
2. *Idea research:* find out all you can about the idea.
3. *Question formulation:* turn the idea into a question that can be answered.
4. *State hypothesis:* guess the answer to the question.
5. *Experimental design:* design an experiment to answer the question.
6. *Experimental activity:* do the experiment.
7. *Observations/results:* watch and record what happens.
8. *Conclusions:* answer your question.
9. *Write-up/reporting:* explain what you did.
10. *Backboard and display preparation:* illustrate what you did.
11. *Display:* show and explain your project at the science fair.

This is an educator's view of the scientific method. This method is specifically designed for experimental types of projects, but some projects do not fit into this model. These are classed as nonexperimental types of projects.

NONEXPERIMENTAL PROJECTS

A nonexperimental project, as the name suggests, is not an experiment but instead involves research into a particular topic through the library or other institutions. An example is the dreaded volcano project. Make a volcano, show how it works, and then use baking soda and vinegar to make it erupt—no experiment takes place here. Many argue that these types of projects have little educational value and should be used only as an entry-level project (Daab 1988; Knapp 1975; McBurney 1978; Stedman 1975). Both the Canada Wide Science Fair and the International Science and Engineering Fair view nonexperimental projects as less valid than experimental ones. On the other hand, Eugene Chiapetta and Barbara Foots see display-type projects in a different light. They point out that not all research is empirical in nature (using numbers and data) and in fact some of the most noted scientists based their work on the scholarship of others (Chiapetta and Foots 1984). Although we could debate this issue, the fact remains that such projects are viewed in a dim light at higher levels and should be avoided by advanced science students or at least by those who wish to compete at higher levels. Still, if your budding scientist is just getting started or is not scientifically motivated or wants to research a particular interest, then such a project would work well. After all, one of the goals is simply participation in the science fair.

THE PROCESS FOR NONEXPERIMENTAL PROJECTS

We need to have a process for a nonexperimental project as well. You'll see that this process is very similar to the one for experimental projects.

1. *Idea generation:* come up with the idea for the project.
2. *Idea research:* find out all you can about your idea.

3. *Question formulation:* turn the idea into a question that can be answered.

4. *Resource searches:* find out as much information on the topic as possible.

5. *Observations/results:* record the information and decide what is useful.

6. *Conclusions:* answer your question.

7. *Write-up/reporting:* explain what you did.

8. *Backboard and display preparation:* illustrate what you did.

9. *Display:* show and explain your project at the science fair.

PLANNING THE PROJECT

A flowchart representation of this process is illustrated at right and in Appendix B. This flowchart can be a planning guide to the entire process. Notice that at least two points in the process involve a sort of check and balance. This check and balance insures that the project is feasible. Both of these places—one after the initial idea research and the other after you start designing your experiment—are as early in the process as they can be.

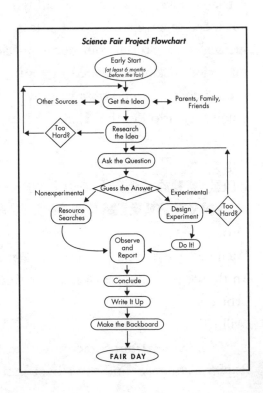

Science Fair Project Flowchart

Both serve as a check that the project is not too difficult to do in the first place or too difficult to get materials for later on. Keep a close eye on your process at those points. Don't waste time—either yours or your child's—on the impossible.

The first step is the idea. We are not talking an epiphany or a little light bulb going off over someone's head, but rather a topic that your child (not necessarily you) would be interested in. This interest is important since it is your child who has to undertake, write up, and present the project. The idea determines whether the project will be experimental or nonexperimental. Where you can get such ideas will be discussed in the next chapter, but any idea or source has worth.

The second step is research. This involves library work, reading, and note taking. Sample note-taking sheets are found in Appendix C and will be discussed later.

The third step is asking a question from that idea. We call that defining the investigation. A question of this type is the starting point of any experiment. We'll not only ask a question, but we'll attempt to answer it as well. This is called a hypothesis. Think of science as asking questions and finding ways to answer them. This is a simplistic explanation but it does the job for our purposes. Properly wording the question will determine how well your child does on the rest of the project.

After asking the question and guessing the answer, the two processes split. For an experimental project, we design the experiment. Designing an experiment depends on what question you asked in the first place. (See how important asking the right question was?) The harder the question you asked, the more difficult the experiment will be to design and carry out.

The next step in our experimental process is actually doing the experiment (nonexperimental projects don't have this step). Now is the time to do what you designed earlier. Most experiments take some time, so it's important to begin this stage as soon as possible. While the experiment is running, someone has to observe what's going on and carefully record those observations, especially in the case of a long-term experiment.

Once we have the results, we have to analyze them and reach a conclusion. What have we learned from the data we have gained? For example, if we have watched plants grow over the past few weeks, we must compare the original value to the daily growth pattern or some such value. This is where the project begins to come together.

Finally we must report our findings and state our conclusions. Students do this in a variety of ways, but the most popular method is through the use of a backboard and a student report. Reporting the findings is just a straightforward writing down of the facts, but the conclusion is another matter entirely. The conclusion is supposed to show the correctness of the hypothesis. Was the hypothesis correct or not, and why?

That's the process as we see it. It is quite involved and must be started early to work well. This type of process can't be thrown together at the last minute. To help both you and your child get a better sense of timing, we have included a little timeline reminder below.

A good science fair project is quite a bit of work and cannot be left until the last minute. Following this timeline ensures that the work is spread out over the longest period and that the students will have plenty of time to do what they need. Some schools actually start their students on the road to a science fair project as early as the first week of school. This isn't a bad idea as long as it is monitored and reported properly. Other schools provide little monitoring and even less support. These schools produce better independent learners but markedly fewer of them. In either case, your child's teacher and school will play

Timeline

Six months before the fair	Choose a topic.
Five months before the fair	Complete initial research.
Four months before the fair	Create the question and begin designing the experiment.
Three months before the fair	Complete experiment's design, gather materials, and set up experiment.
Two months before the fair	Begin running long-term experiments or complete one full run of short-term experiments.
One month before the fair	Finish up all runs of experiments (repeat if necessary); start preparing written report.
Two weeks before the fair	Finish written report; start on backboard and practice sessions.
One week before the fair	Finish backboard and continue with practice sessions, make sure everything needed is available, and make a checklist.
Day before the fair	Pack up everything for transport, make sure it can be set up again, and check off everything on the checklist.

a large role, but support from home is still essential—the schools cannot do it all.

BE SUPPORTIVE

Support is the key word here: involvement but not interference. Students are the ones who have to defend the project to the judges, their teacher, and their peers. The project must be their own work if they are to do a good job in that defense. Parental involvement is an essential part of the learning process for students—not to mention the growing-up process. Yet some parents have been known to do too much for their children's science fair project. If you feel that it's your job as a parent to help, then provide support. Learn the processes your child will have to undertake and provide help at each of the steps. In the end you may end up closer to your children—they may learn from you, but you will also learn from them.

CHAPTER 4
The Idea

The act of choosing a topic is the most difficult part of the science fair process.
(Asimov and Fredericks 1990)

First and foremost on our list was finding an idea. It's first because this is where any science fair project has to start. You will find that for the most part, the project will proceed in the sequence explained in the last chapter. The idea stage is really about two things: first, finding that award-winning (or at least self-satisfying) idea, and second, finding it early enough in the year to get the project done for fair day. Get your kids thinking about the science fair when they start classes for the year. An excellent practice used by some of the better science teachers is to have the students turn in monthly work reports on their projects. This gives the teacher a sense of how well the project is going, and also helps the student keep track of the time passing by. If your child's science teacher is not implementing something like this, you could do it yourself. Home implementation should be just as effective and you would have the flexibility of checking as frequently as you felt necessary.

Quite a few articles about finding an idea have been published in parent magazines and educational journals. The vast majority say this is the toughest part of the science fair project process. John Leibermann (1988, p. 1067) sees finding an idea as "the biggest obstacle to overcome in doing a project" while Bombaugh (1987) suggested that more than 30 percent of any project time would be spent determining the topic. Just what you needed to hear—the first step is also the hardest step! Well, it doesn't need to be. Ideas are available all around us; just encourage your children to ask questions. As little kids they asked questions about everything: "Daddy, where does snow come from? Mommy, why is the ocean blue? Grandpa, why are there wrinkles on your face and why is your hair white?" In each of these questions is a science project just waiting to get out. How many science fair projects can you see in these questions? There's no correct answer—if you can see one science project idea you're doing fine. That's the point: ask questions and do a project to answer them. Using creativity to come up with an idea for a science fair project is the highest form of learning possible—it's also the most difficult and unrealistic. Students need to be inspired, whether it's by something they've read or something they've seen. Nature walks, scientific tours, university visits, or chatting with parents are just some of the ways to find inspiration.

SOURCES

In a recent study at a regional science fair in Canada of where students get their ideas, a number of popular choices were discovered (Barron

1997). It is interesting to note that these sources are also commonly found throughout the science fair educational literature. Some useful sources are:

- ➤ Textbooks
- ➤ Teacher-supplied lists
- ➤ Science magazines
- ➤ Scientific journals
- ➤ The internet
- ➤ Parents
- ➤ TV/Radio
- ➤ Self-creation

Some of these sources are better than others, either because they are more accessible or because they contain more possible ideas. Some schools encourage the use of one source rather than several. For instance, some teachers may send home a list of possible ideas to help their students. Two noted educational researchers, Barry VanDeman and Phillip Parfitt (1985), suggest the use of teacher-introduced topics earlier in the year. This is a very popular technique, both for entry-levels and later grades as well. Many scientific researchers condemn this practice, but there is merit here. Although the answers to most of the questions on such a list have already been determined, recreating an historical experiment is an excellent way to *start* a student on a science fair career. For the more advanced student, taking the idea a step further than what was previously done would also work. This is actually a process of science, working to advance the findings of an existing experiment.

Textbooks have also been condemned by the scientists, but perhaps a bit too hastily. Brian Hansen organized his elementary fair around a list using, "[s]uggested topics . . . from the students' science texts"

(Hansen 1983, p. 10). Pushkin concurs, recommending "that the students look through their textbooks for ideas" (Pushkin 1987, p. 962). A text would allow much the same extension of the idea that a project list would. A student can actually take an experiment from a text and reproduce it (this has little educational value) or he could take that experiment, look at it carefully, discuss it with his teacher—and you, of course—and develop something new and exciting from it. The process here is up to the student and the availability of research materials.

Science magazines and journals offer the best chance of stumbling across a good idea. Experiments are rarely described in enough detail to copy them directly, but ideas are found at almost every turn of the page. Students can either develop their own approach to reprove an existing scientific theory or expand on a theory that already exists. The lack of detailed information on the processes followed will ensure that the student has done much of the work.

Television and radio offer some ideas as well. Science educational shows are an obvious source, but a smattering of science is present in many shows, such as science fiction, mysteries, and medical dramas. How about developing a better way to take fingerprints or using magnets to remove metal splinters? The possibilities are endless—limited only by our access to resources and funding.

The Internet seems to hold the answer to almost everything. You can download educational material by the basketful, everything from reports on the plays of William Shakespeare to a science fair project. The problem is that few of the sites on the web actually police what they are doing. Sample science fair projects are available from several different sites, but that defeats the purpose, which is for students to do the work and through doing the work, learn. Some sites give important organizational information and some sites will help students through the process of developing a project. The trick is to find the sites that are helpful. Appendix D contains a list of sites that will help your child through this process.

This list is far from all the places you can inspire your child to get a science fair idea. Isaac Asimov and Fredericks (1990) suggest that a list of questions that would bring out the student's interests could spark some ideas. Ronald Giese, Julia Cothron, and Richard Rezba suggest "a simple questionnaire asking students to identify their hobbies, part-time jobs, talents, science articles they read, or any science-related interests can help identify topics." As well, books on science tricks, demonstrations, popular magazines, science course-related materials, or lab manuals are other suggestions, once again allowing the students themselves to choose whether to lead or follow in posing a question (Giese, Cothron, and Rezba 1992, p. 32).

CASE STUDIES

For a more in-depth look at the process and the sort of thing that can happen to inspire an idea in a student, the following examples are offered.

What has happened to Erin happens to students all the time on many levels. Interest in *what* your child is doing at school is more important than *how well* your child is doing. Erin has not participated well in the science fair in the past because she hasn't felt it was important, because her parents didn't show her they felt it was important. Perhaps her teachers downplayed the science fair as well. But this year, Erin has an idea she is interested in. Her background includes a fascination in electronics and computers; this interest has sparked the idea that Erin will enter at her local fair. She discussed it with her teacher and now is well on her way.

ERIN

Erin is a grade eleven student at a large high school (more than a thousand students). Although Erin is in the advanced science course, she's had lackluster performances in the science fair over the years. Her heart hasn't been in it, and her parents no longer seem interested in anything except her overall academic standards. So Erin either chooses not to participate or does a project well below her abilities. This year though, Erin has an idea. One day in November, while at her father's work, she noticed the system that the company uses for its security. Erin thought it quite simple, but it was advertised as the best on the market. Erin felt, quite correctly, that she could make a system as good or better. Idea time!

Interest is the key. Many of the science fair pundits will tell you that students must be interested in what they are doing if they are to excel in this area. When all else fails, Erin's interest will hold her to the project and push her beyond the norm. It will also get her started, which is sometimes the largest hurdle to cross. The parent's interest cannot be downplayed here either, Erin's father is supporting her idea by giving her access to schematics of the original system and telling her how it works to the best of his knowledge.

Interest is still the key. Joseph started looking for an idea based on his own interests, and although he needed help articulating what he wanted to do, he finally found an idea that meshed with his interests. Another important point from this example is the teacher's technique: she understands that some students are unable or unwilling to come up with their own project ideas, so she gives out a list. She also understands that some students wish to be creative and go beyond what she offers, so she leaves it open-ended. It should not surprise you that she is an award-winning master teacher.

JOSEPH

Joseph is also a grade eleven student. After attending last year's science fair, Joseph wants to do a project that will do well in this year's fair. His teacher has presented a list of possible project ideas but allows students to come up with their own ideas if they want to. Joseph has thought about an idea for a project for a long time but has not come up with one. All he knows is that he is interested in the environment and helping the earth. Many of the ideas he has come up with were impractical and beyond his abilities and his access to resources. He has talked to his parents about ideas, but his uncle is the one who gave him one. Now Joseph is working on a bioremediation experiment and his uncle is helping him with it. He is interested in the experiment because it is based on his interests, but he needed help coming up with an idea that would fit what he wanted to do.

• • • • • •

WILL

Only for the past year has Will been in a school that does a science fair, and this school has a limited competitive nature. Entry-level projects abound, and teacher lists of ideas are the accepted practice. Students usually do projects that involve proving an existing scientific theory, and Will is no exception. He did this last year and probably will again this year. The teacher list will not be available until January.

Bad habits are forming here, and this is something you have to look out for. Try to get your children thinking about science fair projects well before January. Also, try to get them to go beyond a teacher list or reproving an existing theory when they come up with that idea.

It is important for your child to begin working on real scientific investigations as soon as possible. Your help is invaluable when the teacher is unable or unwilling to go beyond the structured "cookbook" approach.

In the end, using a variety of sources for ideas would be the best bet. Give your child access to as many sources as possible and work with him. Let him tap into his interests so as to move beyond cookbook approaches and the regurgitation of scientific facts. Pushing your child to go beyond what is written for an experiment may be difficult in some cases, but in the end he will feel more ownership of the project.

EXPECTATIONS

What should we expect from our student at this stage? Well, the idea has to be hers, and she needs to start early as well. She needs to identify what she has to do so she can finish the project in plenty of time instead of rushing at the last minute.

What should we expect from our teacher? The science fair actually exists at the teacher's sufferance. Although it is a recommended part of most curricula, it is not exactly required. And just because a teacher is willing to organize a fair, it doesn't mean he will start early enough in the year to be effective. He might also tie your child into a cookbook type of experiment. The first parent-teacher interview session of the year is when you should clear the air about such things. Raise your issues, know your facts, and the teacher will see reason.

What can you be expected to do as parents? Giving your child support at this stage is not that easy. You want it to be her own idea, but you also want her to do something up to her potential. If you feel the idea is not challenging enough, but it is her own idea, what do you do? A wise parent and master teacher once described such a situation, and in the end he let his son do the project his way. Although he didn't do well at the local fair, it was his idea, he felt ownership, and he came back knowing what kind of project would do well. These are all-important lessons, and hard ones to watch happen even when we know they are necessary. Depending on the idea, it may be possible to motivate your child to go further with it than originally intended. If she has trouble getting an idea you can guide her to one at an appropriate level. If you want her to generate her own ideas, give her opportunities to see things around her—visits to the zoo or nature walks might help. Have her list her interests and hobbies; maybe something will suggest itself from these. If all else fails, you can still use the other sources listed—if she goes beyond what is there or modifies the existing experiment in some way, she will have at least taken ownership of it. Finally check up on her progress once the idea has started to take shape. We will discuss this in other sections as well, but this is definitely the place to start checking.

We have just started down the road toward a successful science fair project. As we progress, we will also develop a better understanding of what success in a science fair can mean to both you and your child. To help us get to the end of that road we will be following Erin, Joseph, and Will along the rest of their science fair path. So now that we have an idea, let's turn it into a project question.

CHAPTER 5

From Idea to Project

*. . . students will need guidance at every phase, including
the initial one—selecting a problem to be studied*
(Fields 1987, p. 18)

We are now at the second step of the project development
process—asking the question or defining the investigation.
Once your child has acquired an idea or asked a question, he will
have to convert it into a research problem. There are several routes
to take. The tried and true method is for him to take the idea to his
science teacher and have her convert it into some sort of researchable
project. Of course, depending on the teacher's schedule and the
number of students in the class, the teacher might just convert the
question herself. Being able to change a simple question or idea into
an investigation or research question is a good skill for both parents

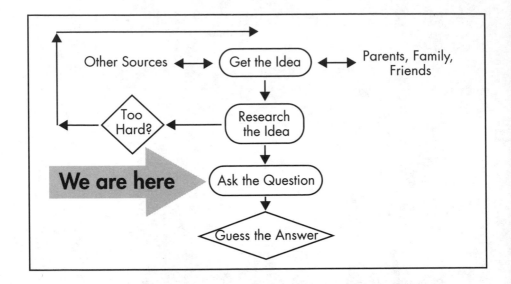

and students to have. Let's begin with a review of those example questions:

Where does snow come from?
Why is the ocean blue?
Why does your face have wrinkles?
Why is your hair white?

Although to you these questions may seem simple, they are questions that any child could ask, and they are questions that science can be used to answer. You may think that because they are already in question form your child can just take one of these ideas and do a project. That's not the case. Another important aspect of science fairs is the use of scientific language. Not only does your child have to answer a question using science, but that question must be phrased using scientific language. Mastery of scientific language requires scientific training, but for our purposes mastery is not needed. Just have your child use the language he's used in science class (if you can remember your own science classes, you can help too). If all else fails, there are thesauruses on the market that deal with just this sort of thing.

So how do we turn an idea into a question? First we need to recognize what elements of science may be present in the idea itself. Then we need to look at what sort of things we can answer and what is beyond our scope. Next we need to word the question in such a way that we can answer it scientifically. Finally, we have to propose an answer to the question or create a hypothesis—and, don't forget, we have to word it using scientific language.

Let's do a quick example of what we mean by converting questions into projects. We'll take our sample questions and translate them into scientific questions using scientific language:

Where does snow come from?	What is the origin of ice crystals in the atmosphere? Why does precipitation change state?
Why is the ocean blue?	How are light and color produced in seawater?
Why are there wrinkles on your face?	How does the skin age?
Why is your hair white?	How are age and hair pigmentation related?

You are not trained to recognize these as projects, but few people are. Science teachers study the philosophy of science to help them do things like this, and years of formal educational training are required for this process to work in a classroom. The only edge most teachers have on parents is the understanding of the structures and the experience of going through it before with many different students.

STEP ONE: KNOW WHAT SCIENCE IS PRESENT IN THE IDEA

We need to know everything we can about our topic before we can go on. Many science teachers use brainstorming to help with this step. Brainstorming involves writing down the main idea and then writing all the things that come to mind from the main idea. In science we use brainstorming to determine everything we do and don't know about an idea. What you do know may form the background to a project while what you don't know may help you form a question for the project itself.

Brainstorming is one step of the process, and for entry level projects this may be the only step needed. But for most projects, after we brainstorm we are going to realize that we know very little about the topic. So we need to do some background research. This has to be done sometime in the process, and by doing it at the beginning we get a head start on our workload as well as help with the question formulation stage. Researching topics has become easier over the years: rather than having to look through mounds of card catalogs at the library we can usually use a computer system to search for information. The Internet is a wonderful place to start a search. At this stage the information we gather does not have to be very in-depth—we are looking for a basic understanding of our topic. Of course, what we find here will be useful later, and we can always return to these sources for more information. One important point that needs to be stressed at this time is the need for extensive note taking and recording of information. On the next page is an example of a note-taking log used by students (it is found in full form in Appendix C).

This log has space for everything you need to record: the topic; the title of the article and the author's name, so we can either refer to

TOPIC:
Source 1
Title:
Author:
Notes:

it later or put it into our references; and finally the notes on the topic. Taking the notes on the topic is the most difficult step here, because we have to take a whole book or article in a magazine and get rid of all the information we do not need and reduce the rest into a few short sentences. At this stage of our process, we only need a little bit of information, enough to help us write our question, but of course we can use these sources later for further background or historical information on our topic.

STEP TWO: KNOW YOUR LIMITATIONS

To do it or not to do it, that is the question. Actually the question is "can we do it or not?" and being able to answer this now will save us much time later. Many checks and balances are in place to insure that we don't waste too much time in the process, and this step is one of those. There is a simple rule here: don't bite off more than you can chew. If your first project question seems like more than you can handle, look at your topic again and see what else you can come up with. If nothing simpler presents itself, it may be time to look for a new topic, but many topics will have more reasonable projects present within them.

STEP THREE: KNOW YOUR SCIENTIFIC LANGUAGE

People in every field—computer programmers, mechanics, and scientists—use some specialized language. The language of science is very important in a science fair project. Of course, this depends on age: if a child is too young for this sort of language, a common-sense approach is fine. Elementary projects, for the most part, expect very little scientific language. But once into the realm of the competitive fair—usually at about grade seven—scientific language becomes important. No matter how well an experiment is carried out, if it is going to succeed at a science fair, it has to have the problems stated correctly in scientific language, the procedure laid out in step-by-step format, and the conclusions written in the same style as the original question. This may seem unfair, but to truly understand what is happening scientifically, the student needs to be able to state it scientifically. To answer a question scientifically we need to ask the question scientifically. Otherwise the process of science we are trying to follow is no different than any other learning process your child will use. This should not be a terrific stumbling block. Your child should have been exposed to this sort of language for years now and should be able to word the statements correctly. Your child's teacher can help at this point. A good scientific thesaurus—found in most public libraries—is another resource you can use. Since you also may need to help, let's prepare you with a little exercise. In each of these paired statements, the first is in common language, the second in scientific language:

Ice is ice.
Ice is water in its solid state.

Rust is water and salt eating metal.
Rust is a chemical reaction.

Soda pop is fizzy.

Trapped gas escapes from a carbonated drink.

STEP FOUR: KNOW THE ANSWER (I THINK?)

Yes, we're actually suggesting that you have some idea of what the answer to your main project question is. Sort of strange isn't it? We're asking a question to which we believe we know the answer, so you may wonder why we are asking the question in the first place? Well, we have to *prove* whether or not we are right. This is what a hypothesis is all about: it is an educated guess about the answer to the question that we asked. We then test the question to see if we are right about the answer. The important thing to point out here is the term "educated guess." Many of the experiments suggested by a teacher have already been done, so their answers are already known. Even if you actually know the answer, to make an educated guess you have to explain *why* you think it's going to happen. Just saying, "It's the answer because I know it's the answer," or "I was told it was the answer," wouldn't cut it. Your child will have to justify the answer to the question in the project write-up, so be prepared.

EXAMPLES

We return to our original, seemingly simple, questions. The first was: "Where does snow come from?" You might think that this question could be answered as it is, and for an entry-level project in the elementary grades (fourth to sixth) this might be true. But why settle for this? Working through each step of our process, the first is to recognize what elements of science may be present in the idea itself. Brainstorming about snow may lead us to what is actually happening: water forms on small dust particles in the atmosphere, they get larger and start to fall to the earth, and when the air temperature gets low enough, these water droplets freeze. If we are unsure about the process, we can use the library or other resources to learn more. Once we have established what is happening scientifically, it's time to move on to the next step in our process: is this experiment possible with the resources at hand? Many elementary science texts and encyclopedias cover the process of ice-crystal formulation in clouds in minute detail. So this step can be accomplished by reviewing the literature and text resources that you have. It's possible to grow example crystals at home for a minimal cost, so the cost of resources or the availability of materials isn't an issue. At this point of the process you should determine if the project is experimental or nonexperimental. If the entire project involves library research, then it's usually a nonexperimental project. If some experimentation is taking place, even if the answer is already known, then it's an experimental project. In this example, even with growing crystals at home for display purposes, the project is a nonexperimental project. On to the next step.

Now is the time to put our question into scientific terms. In this particular case we should not find it too difficult. The question is short and simple, so even with scientific language the question should remain short and simple. "Where does snow come from?" turns very easily into "What is the origin of ice crystals in the atmosphere?" or "Why does precipitation change state?" All we have done in the first instance is rephrase "where . . . come from" into "origin," and then use the scientific name for snow. In the second instance we have looked at snow as a process: precipitation, and then looked at what it was doing: changing state from liquid to solid. This is where our background research and a scientific thesaurus come in handy.

We have a topic, an understanding of background behind the topic, a scientific question, and all we need now is an answer. It is time for that educated guess. From our background research, we realize that a drop in temperature below 32 degrees Fahrenheit (or 0 degrees Celsius in scientific circles) has caused a change in state, so this is what we propose as our answer. The rest of the research that we undertake will be to find out if this hypothesis is correct.

This is a good place to discuss the scope of a project. We have already looked at experimental and nonexperimental projects, now let's consider how far we should go. We can grow crystals at home, we can find information about snow forming, we can write it all up, but we are still dealing with a nonexperimental project. We could take this project as step further and try to re-create atmospheric conditions that would produce snow and then vary aspects of those conditions, which would then be an experiment, but we have to be realistic. As an experiment, this project would require access to materials and facilities beyond what we should expect, so we better keep this project's scope at the nonexperimental level. You'll find that in almost all cases you can reduce what you do with a project to make it nonexperimental or increase what you do with a project to make it experimental. That gives you some leeway in dealing with different age and skill levels among children.

Let's take a look at our next question: "Why is the ocean blue?" Once again, for an entry-level project, this question would be fine. To go beyond this we have to use our process again, starting with discovering what elements of science are present here. In this seemingly simple question we are dealing with light, color, seawater, and how each of them interact. If you have a scientific background, you would know that white light is composed of all the colors, and that it separates into those colors when entering or leaving a substance thicker than air. If you don't have a scientific background, you could have found this information from the library. You could also have learned that seawater is not truly blue, but mostly clear, like tap water. Finally, you would have learned that seawater actually reflects the color of the sky, which is blue, and this is why the ocean seems to be blue.

It's time for scientific language once again. You'll notice that the level of this project is much the same as the last, with nothing but library research involved, so this is also a nonexperimental project. Let's try our skills on this question, which like the last one is also very short and simple. "Why is the ocean blue?" translates into "How are light and color produced in seawater?" Obviously we're dealing with color, which is both a common term and a scientific one; we're also dealing with light, and with the ocean, which consists of seawater. Our translation leads nicely into our next task, that of answering this question. In this case our answer could be something like the sky is blue and that color reflecting onto the ocean causes the ocean to look blue as well. This is of course the simple answer; a more complex answer would be expected within the project. We would discuss everything we learned about how the sky became blue through diffusion and refraction of light, how reflection works, and what are the properties of seawater—and the oceans in particular—that cause this reflection.

Our next question, "Why are there wrinkles on your face?" should unfold much like the previous two. First visit the library and find out that the wrinkles are a function of the skin's aging process and are a result of a lack of skin cell regeneration and moisture in the skin. Although an experimental project is well beyond the skills of most students, the information on all of this science is readily available. Putting our question into scientific language yields, "How does the skin age?" Our last step, posing an answer to our question, could easily be something along the lines of: as we age the skin dries out and pulls together, causing wrinkles.

Finally, we look at the last of our seemingly innocent questions: "Why is your hair white?" Up until now our questions have yielded only nonexperimental projects, but this question has the possibility of an experiment in it. We could describe the hair aging process by going to the library, but it doesn't take very much extension of this topic question to give us an experimental project. The materials are readily available: white hair from the student's family or friends. We start in the usual place, asking what elements of science are present in this question. Human hair, color, and pigmentation seem to be the largest contenders. Our research will have shown us that white hair is not actually a hair color but an absence of hair color, and that hair color has to do with pigmentation in hair. It should also have shown us that this loss of the ability to produce color in hair is a genetic matter that varies from person to person and family to family. Now we have a problem—this project was supposed to be an experimental one but if we continue with this line of research it won't take long to get beyond our capabilities. This is a good example of when to refine your question into a workable project. Subtopics available from this question include questions about the strength of hair and how that strength changes with age or with a change in color. All we need for this to be an experiment is to have something we can vary and test. A subtopic of this question—hair strength—will create a project that is well within our grasp. If we cannot study the hair color as a function of age, why not study hair strength? So we refine our topic to: "How strong is your hair?"

Once again research is in order, which should reveal that the science in this question involves the tensile strength of hair and the age of the individual from which the hair was taken. Second, we look at converting our question into scientific language, and with our new terms it seems pretty simple. Testing the tensile strength of hair versus the age of the hair host, or something to that effect, will work just fine. You can see that an experiment into a topic such as this is very possible. We now have only one step remaining: our hypothesis. The problem with this step is that because it's an experimental project, our answer should be a little less researched but still have a justification. What this means is that we don't want to be obvious about doing an experiment for which the answer is already known, but we still need that answer, so our answer statement becomes something like this: "I believe that hair becomes less strong as you age because of less calcium in the body, because the color has gone out of it, and because of the frailty of other parts of the body as we age." It's a little long-winded, but it gives the justification you need and suggests what you expect to get out of your experiment.

Let's revisit our three students and see how they are doing.

ERIN

Erin, our grade eleven student, is still working on her computer alarm system. Although not an experiment in the strictest sense, it has some grounding in that area. This project would fall into the technology category at the ISEF and will be an "innovation" at Erin's local fair. In her own way, Erin followed the process set down here: she did her background research, identified her major technological problems, and put down on paper what it was she felt she could do. Her teacher has been as supportive as possible, but expertise in these areas within the school system is limited. Nevertheless, the project is unfolding with schematic diagrams and small-scale test devices; she soon will be ready for her actual "experiment."

• • • • • •

JOSEPH

Joseph, our other eleventh grader, is also well on his way to identifying what his question about bioremediation is actually going to be, and he has made many visits to his uncle's. He hasn't had to rely on the library as much as Erin has, since his uncle has loaned him many resources. Still the background material he's found has shown him that very little has been done in this area at his level. After only a few weeks of thinking and reading his uncle's texts did his topic come out: "Bioremediation of oil-soaked soil—a test of cleansers." The access he has to his uncle and the resources of his uncle's workplace make this a possible prizewinner.

• • • • • •

WILL

Meanwhile Will, our entry-level student, has finally received the teacher's idea list. He has had three weeks to look it over and he's decided to do a project involving sound and plants. His teacher has devoted several class periods to the science fair, so he's had class time to get his questions answered and do some of his necessary background research. In this case the teacher helped him with his question, which is: "How do different volumes of sound affect plant growth?"

All three of these cases illustrate different points on a spectrum of project help. Erin is mostly going her own way, beyond the help her father can give her with schematics, because of a lack of resources in this area within her school. Joseph is getting some help from his uncle, but mostly in the form of access to resources and resource people, so he too is developing his project mostly on his own. Will's case takes us to the other end of the spectrum with the teacher supplying the idea, the time and place for research, and the question to be answered. The important thing to note here is that learning is taking place in all three cases. Erin is probably learning the most, while Joseph, through his uncle's mentoring, is a close second. Will is also learning, but what he is learning is process, and for an entry-level participant this is fine. We hope you see how inappropriate that kind of teacher (or parent) interference would be at the later grade levels.

So now that we've found a topic, asked a question, and hypothesized a solution, it's time to design an experiment (if necessary) to answer that question.

Designing the Experiment

Allowing pupils to design their own investigations
will offer opportunities for them to develop
many cognitive skills.
(Hudson 1994, p. 100)

A s you can see from the flowchart on the next page, we have now
come to a parting of the ways. The projects that are nonexperimental in nature take a different fork on this road than those that are
experimental. Chapter 9 will take us further along that fork. If we
are doing an experimental project, it's time to design our experiment.

An experimental design consists of more than just the idea. It also
consists of the apparatus, the expected outcome, and the actual experimental activity. For now, we're going to concern ourselves with the

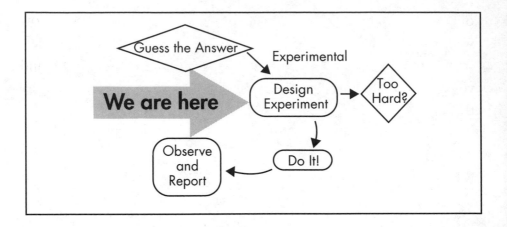

apparatus and the activity. The apparatus is what we are going to do our experiment with—be it seeds or radioactive material. The interesting thing about the apparatus is that most of it will come directly out of the experimental design. Here is where defining our investigation—turning our idea into a question that can be answered—plays its biggest role. What you do from here on depends on what question you asked. There is a process to designing an experiment just as there was a process in the previous sections; finding it is the difficult part. Many educators feel that the process of experimental design is a figment of our imaginations, and many of them would be right—although not in the way they think. Designing an experiment requires imagination and creativity. We have to look at our question and figure out a way to answer it. We can break down this part of the process into easily followed steps, just as we did before. The first step would be to take a good, hard look at our question and try to determine exactly what is involved in answering it. What part of our question is testable? What things do we need to affect or change in the course of the experiment? The second step is identifying what we are *not* going to test or change in the course of the experiment. The third step involves identifying the materials we will need for the previous steps and then obtaining them. Finally, the fourth step is to write it all down. Now we'll explain each of these steps fully so that there is no misunderstanding. Experimental design is an important part of the process and you should make sure you understand it thoroughly before you start the actual design.

STEP ONE: WHAT TO DO

From a scientist's point of view, this step involves determining what is testable within the question. We need to do about the same thing. The problem is recognizing what is testable. We are trying to identify which things from our question we can test or change to help us answer that question. These are called variables, because through the course of the experiment we will vary these things.

A simple example involves looking at how plants grow. What things (variables) affect the growth of plants? Well, let's first look at everything involved in the growth of a plant and then break it down into the things (variables) we can test. We know that air, soil, water, and sunlight affect the growth of plants. We can test any of these things. Changing the air quality, changing the amount of sunlight, changing the amount and type of water, or changing the type of soil are all good ideas for doing an experiment on plants. One thing you should remember though is that a good experiment involves changing only one thing at a time.

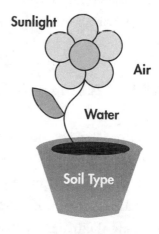

Sunlight

Air

Water

Soil Type

This is, of course, a simplification of the variables involving plant growth. In reality, each of these variables can be broken down into smaller variables, and some more complex variables such as fertilization have been left out. However, this gives us a good place to start.

STEP TWO: WHAT NOT TO DO

We have identified the things we can change in an experiment, so now let's move to step two and identify the things we don't want to change in the experiment. This will be covered in more detail in the next chapter, but for now let's just call this controlling variables. Controlling variables is an important part of the experimental design and carrying out stage. This may prove the most difficult step, though your child should have had at least some exposure to this concept before now. What things do we need to keep constant in our experiment so that we are actually testing what we're trying to test? In our plant example, if we want to test different soil types on plants, we would need to keep the other variables—water, sunlight, and air quality—constant for all of our plant samples.

STEP THREE: YOU NEED WHAT?!

Sooner or later you are going to need the actual physical parts of your experiment. Scientists and science teachers use many different terms for this, but the word "materials" will work. "Apparatus" is another common term. Basically, it's the things we need to do our experiment with. Obtaining the materials should not be a difficult part of your experimental design if you have been careful with the other parts of the process.

The materials don't necessarily have to be expensive or difficult to obtain. You can sometimes substitute common items for more expensive materials. Household items can be used in many cases. If your child wants to test acids and bases, household goods such as vinegar and baking soda will come in handy. Old television sets are a treasure trove for such things as resistors and capacitors (make sure to unplug the TV before opening it up!). Even lasers can be had for a reasonable price if you know where to look—a discarded item-checking system has a laser in it, or use one of those irritating laser pointers.

Of course some materials are just impossible for a school-age child to obtain. Not many parents have access to proton accelerators, or liquid nitrogen, or even animal blood. If you can't find—or afford—the essential materials, you may realize at this point that doing the experiment is impossible. We had hoped to recognize the viability of the experiment well before this, but at least this part of the process will prove as another checkpoint so that all is not lost, and no more time is wasted.

One solution to the problem of inaccessible materials is mentorship programs, which are offered by some science fair jurisdictions. These programs help by pairing a child (or a number of children) with a credible scientist. The scientist provides access to materials and laboratories that the students wouldn't normally have and also offers sage advice to help the student along the science fair project road. This approach may solve your materials problem, but only if your region offers this service, and only if you are within range of the facilities. We urge you to question your child's science teacher about such things or contact your regional/state science fair organizers.

In Ohio there's such a mentorship program, involving local scientists. Fifth grade science students are paired with researchers from many institutions in their general area, allowing the students to investigate with the researchers "a wide variety of topics under the guidance of interested community members" (DeBruin et al. 1993). Essentially the resource person becomes the primary source of the research, with scheduled meetings taking place before the fair so that all background materials and research can be checked and validated.

Mentorships can be less formal as well. Joseph, for instance, has an informal mentor-type relationship with his uncle. In fact, on some levels, the relationship that many students have with their teachers concerning the science fair is a mentor-type relationship.

STEP FOUR: YOU DID WHAT?!

We cannot overemphasize the importance of good note-taking and reporting. Writing down the experimental procedure you have undertaken is important now and also later in the process. Students in a science fair are much like detectives; they have to write down everything they hear, see, and experience. This cannot be stressed enough—keeping good notes is an integral part of having a good science fair project, and it's important for many parts of the process. One important fact about science experiments in general—and science fair projects in particular—is that their validity is a function of their repeatability. This is not as complex as it sounds. Here's an example you may have heard about in the news. A few years ago, a group of scientists claimed they had discovered a way of creating a fusion reaction in near-room-temperature conditions. Up until that point, fusion reactions were thought to happen only at extremely high temperatures, so this was a terrific breakthrough. The problem was that when other scientists copied the experiment, they couldn't get the same results as the original group did, so they thought that the original experiment's conclusions were not valid. To this day the debate between those who believe in cold fusion and those who do not still continues in the scientific research world. Validity and repeatability are linked within all scientific research, so you have to keep good notes on your procedure to make sure the judges think your experiment is at least capable of being valid.

EXAMPLES

Now let's practice a little with the questions we have been using as examples. Remember from the last chapter that we had decided to undertake some of these questions as nonexperimental projects. One of the reasons we did this was because we felt that developing an experiment for those questions would be too hard. With luck this will become apparent when we follow through on the steps.

"Where does snow come from?" was our original question, which we converted to "What is the origin of ice crystals in the atmosphere?" Step one is to figure out what are the variables within this question. We have identified the variables—temperature, the amount of moisture in the air, and air pressure—from our background research, so we can decide what we want to test after looking at these variables. As an example, and keeping with our stated hypothesis for this question, let us say that we are going to vary temperature and test what effect this has on ice-crystal formulation. Remember that we suggested that it was the temperature dropping below 32 degrees Fahrenheit (0 degrees Celsius) that caused the ice crystals to form. Moving onto step two, controlling variables, we know that if we are going to vary temperature, then we need to keep both air pressure and the amount of moisture in the air constant. We also may have to keep other far reaching and not easily recognizable variables constant, such as the time of day of the experiment, the amount of sunshine, and the day of the experiment itself. You'll notice that the scope of this project is quickly getting out of hand. This trend continues as we determine what materials we need and try to obtain them. We would need a large atmospheric lab to undertake this type of experiment. Even using computer models, as some meteorologists do, would require heavy computer time on a very sophisticated computer. The last and deciding factor is that all the information you need to answer this question is easily and readily available in most science texts and school libraries. Adding up all these factors led us to the conclusion that we should treat this idea as a nonexperimental project.

The second question that we identified earlier was "Why is the ocean blue?" We converted this question to "How are light and color produced in seawater?" Within the second question we have to determine what factors determine the color of the ocean and attempt to prove or disprove our hypothesis that a major proportion of this blue color comes from the reflection of the blue sky onto the ocean's surface. Our background research into this topic would have revealed that depth, the amount and type of plankton in the water, light intensity, the water's salinity (amount of salt), and the blue color of the sky would be the major variables to consider. Obviously it's the sky, and its color's effect on the color of the ocean, that we will want to vary.

Step two is the simple matter of deciding what we will not vary; in this case, everything except the sky's color would be appropriate. Step three has us scrounging for materials, which will involve a scale model, glassware for holding water, the seawater itself (if you live inland this may prove a problem), and a device for measuring light color and intensity. This last item on our shopping list may be the most difficult to obtain. Although any commercially available light meter for photography allows us to measure light intensity, to measure light color we will need a hand-held spectrometer, and finding that is not going to be easy.

A simple way to do this project is to re-create a section of the ocean and sky in model form. If you are able to keep outside light away from the model, you may see the ocean being blue. Another way is to just take a glass of seawater, isolate that seawater from the blue sky, discuss its lack of color, and then show pictures of the ocean and its blue color. This would serve to support your hypothesis, but not enough to make it worthwhile. The difficulty with this experiment is that supporting what we know is true is almost impossible experimentally. Taking what we know is true from a text or library book is much easier. Once again the scope of the project and our available skills determine whether our project is experimental or nonexperimental.

Our third question was: "Why are their wrinkles on your face?" which we translated to "How does the skin age?" A study of the skin's aging process is not something to be taken lightly in an experimental sense. For starters, experimenting on humans poses a lot of problems, and under the rules of the International Science and Engineering Fair this means a lot of hoops to jump through. Second, the scope of such a project would be more than even an above-average student could handle. For completeness, and to give us a better sense of the project if we decide to do it nonexperimentally, let's look at our possible testable variables. We have discovered that wrinkles are a function of the skin's aging process and are a result of a lack of skin cell regeneration and moisture in the skin. We also discovered that the variables that may cause this condition are very numerous and may include genetics, exposure to sunlight, the amount and type of soap used daily, etc. Any of these could be a testable variable, but the most easily tested is the amount and type of soap used. All the other variables—genetics, the amount of sunlight, and any preexisting skin conditions—would have to remain constant, which means we would have to use the same person throughout the experiment. Now it is time to look for materials, and on the top of our list is a good quality microscope. We are going to have to look at skin cell samples from a person's face at different times of the day, so we are also going to need a person. Finally, we would need some commercially available brands of soap to test.

Now we look at our final question, which has been changed from: "Why is your hair white?" to "How strong is your hair?" which we then translated into "How does the tensile strength of hair relate to the age of the hair host?"

By refining our question, we have an opportunity to do an experiment rather than just a nonexperimental type of project, which the rest of our questions turned out to be. So let's take a good look at this new question to make sure we know exactly what is testable. Since we took so much care with our question, what is testable falls right out. Tensile strength of hair is our testable variable, while the age of the person who owns the hair is another variable we will need to change. The length of the hair, perhaps the cross-sectional area, and whether it is colored or not would be the other variables in this experiment. Those variables would be our controllable variables. Our materials are readily available, just pick the hair of a few generations of your family. By using one family, we make family lineage another variable we can keep constant. Good record keeping is very important here—we need to track the origin of each hair sample. An example of a good experimental design worksheet is shown below. Note the spaces for identifying all your variables and then separating them into controlled and uncontrolled. A worksheet of this type is necessary to ensure that nothing gets missed and that the experiment can be repeated.

Variables	Controlled Variables	Uncontrolled Variables
Topic:		
Question:		
Hypothesis:		
Apparatus:		

So what do we have to do for our experimental design? We need a number of hairs from each member of our family, and we need some way to test the tensile strength. Brainstorming is one effective way of coming up with tests of this type; another way is to use known techniques found in science texts or magazines. The test we think is the most appropriate would be to anchor one end of the hair on a hanging board of some sort and then add weights to it in small increments until the hair breaks. We then take the weight present on the hair just before it broke (remember to write everything down!) and this weight corresponds to the tensile strength of the hair. We'll get into more detail of the experiment when we carry it out in the next chapter.

CASE STUDIES

ERIN

Erin has progressed about as far as she can by herself. She has identified her experiment, which is a test of her security system. She realizes that the variables she has to control are the sensing devices themselves, the physical space they are in, and the cost of the equipment. Cost is not usually a variable in science fair projects, but when you are dealing with engineering and computer technology innovations, cost can be an important factor in determining whether the project design has merit. Her innovation involves using the existing system's motion detectors and her father's office area, but moving the control panel to the computer on her father's desk. The programming she's done will theoretically allow some of the system to be turned off while keeping most of the building secure. In case of a break-in, it will send an automated phone message to the manager and to the police. She has received permission (after many letters and phone calls) to test her system with the police dispatcher. (Many government agencies, such as police and fire departments, research labs, and universities support the science fair movement by providing technical support for student projects.) All that remains is for a running test to be undertaken. There are several hurdles that Erin must overcome before this can be a reality though, and these will be discussed in the next chapter.

• • • • • •

JOSEPH

After talking things over with his uncle and using one of his texts, Joseph has identified his variables as the type of oil, the amount of oil in the soil, the soil type, and the enzyme (a protein that helps speed up the natural breakdown of various substances) used to break down the oil. That enzyme is the only difficult thing on his list of materials, but his uncle can provide that. The variables he has chosen to control are the enzyme and the oil type. Among the other materials he will be using are large glass beakers, different types of soils, different types of oils, and a graduated cylinder (a tall beaker with volume measurements on the side) for measuring the oil.

• • • • • •

WILL

Will chose an experiment from the list his teacher has given him and is now the proud owner of the question, "Does sound affect plant growth?" He's been busy identifying his variables (with his teacher's help), which are type of sound, volume or loudness, plant type, soil, watering, and plant age. He has determined that what he wants to test is actually what volume of a constant sound a plant can stand. His mother told him that some people talk to their plants, and he has incorporated this into his experiment. You can see that many teacher- or parent-provided ideas can be expanded into very good science fair projects.

EXPECTATIONS

What should we expect from our student at this stage? Your child got a topic, converted it into a question, and designed an experiment to answer that question. This has been a lot of work, but it is the only way to truly learn the process. This work needs to be completed early enough in the process so that there is time to do the experiment at least once and maybe twice before the fair date. Keeping good records of the process is very important as well.

What should we expect from our student's teacher at this point? The teacher may need to mentor the students on different aspects of their project, help the students with the question, and help in the design process as well. The teacher can be a valuable resource for students to draw on.

What are you expected to do as parents to aid your child at this stage? Think of yourself as big brother or sister at this point. Give your child some advice, some support, and an occasional ride to the library. Your child may find it difficult to make that initial contact with a family friend or scientist uncle and this is where you can make a connection—for your child and with your child. Don't forget our cardinal rule though; be informed, be supportive, but don't interfere.

Doing the Experiment

The feeling of accomplishment derived from
carefully planned and executed experiments is . . .
[an] important source of motivation.
(Liebermann 1988, p. 1067)

S ome educators would argue that all the preparation we have done
up to this point is for this part of our process—doing the experiment. Yet, when we look at our flowchart we can see that we are progressing along one leg of the science fair project journey, the leg that represents the experimental project. Before we designed our experiment, the paths were the same, but here they diverge.

Within the school system there are very few opportunities for students to actually undertake experiments that they themselves have designed. This makes the science fair project even more important, as it may be the only chance that many students have for undertaking their own experiments. Students need to be motivated, and this part of the project—getting into it, getting their hands dirty, and taking pride in what they accomplished—forms the basis of that motivation (Leiberman 1988).

An experiment needs certain things if it is to work, and these things can be set up in some sort of order. You have to do a number of things before you actually get a chance to start your experiment. Some of them are very obvious but need to be mentioned so that they will not be overlooked, while others are not as obvious.

SCROUNGING

First on our list is to obtain the materials. As we noted before, many materials can be scrounged from a variety of common, everyday sources. You can find transformers, resistors, capacitors, and all sorts of other electronic components in an old TV. If you do not have any old TVs or appliances hanging around, try your local repair shop. If you are looking for material for a model or display, try some of those old train sets or make your own landscape using papier mâché. Literally thousands of project materials can be found around the house including chemical substitutions like mild acids (vinegar) or bases (tartar sauce), sodium bicarbonate (baking soda), citric acid (orange juice), and so forth. Your child's teacher may be able to suggest substitutions for many expensive chemicals. In some instances, children can borrow equipment and/or chemicals from their school labs, so it is very important to have them check with their teacher at this stage.

GIVE IT SOME SPACE

Second, the project needs a space of its own. Obviously, it is not going to move into its own apartment, but it does need an area of the house (or garage or basement or wherever) where it can remain undisturbed during the experiment. You don't want to move apparatus in the middle of an experiment. Any change in position can cause massive errors and throw off your child's results. So find a corner that your child can have for a few weeks. Just remember that it has to be a place where the child can set up what is needed, can get easy and frequent access to, and can be reasonably sure that the experiment will not get contaminated by the surroundings (sometimes easier said than done!). A last resort is asking the teacher to find a space at the school. The only problem with that is accessing the experiment during the weekend. Some teachers are willing to make forays into the school on the weekend and some are not—find out which one your teacher is before considering this approach.

CONTROLS, VARIABLES, AND CONTROLLING VARIABLES

We briefly touched on controlling variables in the last chapter, when we talked about experimental design. Basically, controlling a variable means making sure that it does not change. Our testable variable is the one we want to change, but the controlled variables are ones we have to keep the same. Some experiments may use something called a control. This is a little different, because in this case we may run the whole experiment without doing anything or changing anything, just to see what happens when nothing is tested. This is our control or baseline result. For example, let's look at our plant experiment from the last chapter. We want to vary the amount of sunlight on a number of plants and record the difference, so we have to ensure that all the other variables (air, soil, and water) remain the same—that they are controlled. But it is also a good idea to grow one plant in the sun with all the other variables the same as a control. Then we can use this growth to compare to our other plants to show there is a difference from the baseline result. Controls are often used in tests of new medicines. One group is given the drug, one group is not, and one group may be given a placebo (a pill that does nothing). The medical progress of the group that got the drug will be compared to the other two groups to see if the drug had an effect. The other two groups are called control groups.

CONTROLLED SPACE

Some projects need space that can actually be controlled, and this might cause a problem with finding the project some space. The materials have to remain undisturbed by outside influences during the time of the project. For example, if your testable variable was the amount of sunlight present in the experiment, some plants would have to be given more sunlight than others. This may mean that the project can't be put in some completely out-of-the-way place such as a basement. Believe it or not, even after all the other checks and balances that you have gone through, this issue may cause a reevaluation of at least your variables but possibly your whole question. Fortunately, substitutions are available for this type of experiment as well. For example, where the sun is not available, UV lighting could be used. Where the variations outside may cause problems, a miniature greenhouse could be constructed. Thinking clearly and not getting excited (which is easier said than done for your child) can solve most of the problems you may run into with the location of your project.

CHECK THE APPARATUS

The next step is to lay out the apparatus (materials) and check to see that everything is there. This might seem like common sense, but when it's time to do the experiment and you discover that something is missing, you will wish you had double-checked. Good science teachers will tell you that before they do a class experiment, they always set out all the apparatus (materials) they have listed and make sure that everything needed by the students is there. A science fair project should be no different.

RECORDING SPACE

The project is chosen, the materials are obtained, and now it is time to think about record keeping. A designated space for record keeping is not only a good practice for neatness and organization, but it is also important scientifically. This space needs to be situated so that your child can see everything in the apparatus clearly, there has to be room for measuring and recording, and finally it needs to be someplace where it doesn't contaminate or interfere with the experiment. Keeping a daily log is essential for some experiments, and if a log is required, having a designated space will ensure that it is kept safe and may even remind your child to write in it. Appropriate record keeping will be discussed in the next chapter.

RECORDING THE APPARATUS

It is very important to keep good records for your project. We'll discuss recording the data in the next chapter, but your child should make diagrams of the apparatus and setup as well. There are several good reasons for this. For one thing, something may be dislodged in the apparatus and your child may need to refer to the diagram to be able to return the piece to its proper place. The second reason has a scientific basis—repeatability. An experiment may have to be undertaken more than once to get a desired result. Having a number of trials of a particular experiment with similar results makes the results more statistically valid. The third reason has to do with the display portion of the science fair project process. In chapter 1 we discussed how large a display can be, what can safely be displayed, and other rules that need to be followed when entering a science fair. If the apparatus is too big or uses some sort of substance that is banned at science fairs, then it cannot be displayed, and your child will need to display pictures of the apparatus. This can be accomplished by taking pictures and/or drawing diagrams (using both is best). Remember that sometimes rules change, and what you thought may have been an acceptable substance may now not be allowed—and finding this out as you and your child enter the fair site is a big problem. A picture is worth a thousand words, so having some pictures and diagrams certainly won't hurt your child's chances.

CHECKING IT TWICE

Well, everything is set up, all the materials are accounted for and laid out, a space has been set aside for the experiment and the recording, diagrams have been drawn, pictures taken so that everything is ready for your child's experiment. It is now time to actually do the experiment. Make sure to check on a few things first. First, make sure that your child has started the experimental procedure early enough that it can be repeated or started over if necessary. Even the best-planned experiments might not work for some uncontrollable reason the first time. And don't forget that some experiments need to be repeated to make them more valid. The second thing to remember is to take a good look around for possible sources of errors. Any experiment may require some explanation as to why it didn't work exactly as planned and it is always nice to know some of these possible sources in advance. Last but not least, make sure that you and your child have checked the experiment carefully for safety hazards—and not only the apparatus itself but also the area around it. Now you are ready to begin your experiment.

EXAMPLES

We have one project idea that we knew we could do experimentally: "Testing the tensile strength of hair versus the age of the hair host," but one of our other projects also has a chance to be experimental in nature. That project was: "A study of the skin's aging process." We realized that the project as originally designed was beyond the abilities of most students, but what if we changed it, took it a bit further, and set up some sort of consumer project? A consumer project is another form of a project that is used within the science fair, and whether or not it is experimental is open to debate. A consumer project tests commercially available products, such as different brands of ketchup or toilet paper or paper towels, just like you would see on TV. Actual consumer tests control variables and may even have control groups, but as this falls more within a test of a product, some argue that it is not an experiment in the true sense. We will include consumer projects in with the experimental projects although we recognize that it is sort of a category of its own. So if we choose to do this experiment as a consumer project, we could do something like test two kinds of soap on a person's face, keeping all other things constant and equal. We could then compare the moisture content in the skin as well as how well the soap cleaned. To do this we will need several things: brands of soap to compare, volunteers (a nice way to say guinea pigs), a microscope, and sterile wipes to test how well the soaps cleaned and moisturized. The volunteers are needed to try several brands of soap over a period of time, each time with one brand used on one side of the face and another brand used on the other side. In between each test they should prepare their faces by washing with their own soap for a couple of days or so, which may keep outside contaminants from interfering with the test. The same two or three volunteers should be used throughout (this experiment has family members written all over it). The wipes are used to take samples from the person's face soon after washing and the microscope is used to study the wipes.

As this experiment is going to take several weeks or months, the apparatus for it, namely the soaps and the wipes, need to be mobile. The observation position, though, should be stationary. We need a way to transport the soap and wipes to a proper location for use and then to where our microscope is. The microscope and record book should be close together so that results can be quickly copied down. Other variables, and there are a lot of them, include the way the volunteer washes her face, the time of day, whether the washing takes place in the shower or out, humidity of the day, time of the year, and so on. Some of these variables we'll just have to record, because to attempt to control them all would require a large plastic dome around the house! So now this experiment is ready to go.

We interrupted our experiment on "Tensile strengths of hair" so let's go back and take a look at how it could unfold. First, we need to scrounge some materials for our project. If we look at our experimental design from the last chapter, we see that we were planning to attach weights to the hair and add weights until the hair broke. The weight the hair holds before it breaks will (we think) represent the strength of the hair. We're going to test hairs from people of different ages in our family. So, we need some kind of frame from which to suspend the hair and weights, we need some measured weights, and we need hairs from several generations of our family. The frame can be put together easily with some old scraps of lumber or by using the underside of a table or desk. The only drawback is that we cannot take the setup to the fair to show everyone exactly how we did it, but we could take pictures. Measured weights are weights that have a known value, and they would have to be relatively small so that they could be suspended from a strand of hair and not break it right away. Many schools have weight sets that can be borrowed for a weekend. Otherwise, you could take some washers, weigh them at school and record the weight of each one, and then use them to suspend from the hair. Finally we need the hair. While your child might think it would be fun to pluck hair from family members, it might be better to use a clean brush and take hair samples from it after having the family members brush their hair.

We have scrounged our materials, so now it is time to find a place to put it all. The space assigned to this project has to be undisturbed for the time it takes to accomplish the experiment. Even with twelve or more hair samples (assuming three samples from four family members), the experimental part of this project can and should be accomplished in a single afternoon. We'll talk about the observation part of the experiment in the next chapter, but for now let's just say that completing the experiment in one afternoon will help reduce environmental variables caused by different temperatures and such. Given the small amount of time needed for this project, finding a good space for it should not be difficult.

Once we have the space, we need to lay out our apparatus to make sure we have everything we need. Setting up for a dry run is a good idea that will tell us right away if we have missed anything important. And you know what? We have! We have not determined a way to attach the hair to the weights. We could tie the weights to the hair, but the knot introduces a factor that we do not want, so we need some other way. In the end, after testing various possibilities, we came up with hair clips. It's important to understand that in doing an experiment, sometimes you have to set things up by trial and error. Use different materials for certain parts of your experiment to see if they will work, and if they don't, move onto something different. The hair clips were decided upon after trying and rejecting paper clips, glue, tape, fishhooks, and other possibilities. The hair clip connects the hair to a string that is tied to the weight. Each new weight is tied to a string and then reconnected to the hair carefully to avoid pulling on it. We discounted the other possibilities because we were dealing with small weights, which can be affected by variations such as knots and glue and such. Remember, we have to be careful not to introduce other, uncontrolled variables into our experiment.

Now we need a place to record our findings. In this case, having a space right next to our experiment shouldn't be a problem. Finally, we'll set up some sort of data record chart, which we'll discuss in the next chapter. It seems that all is ready and we can do our experiment. Keep reading to find out how it went.

CHAPTER 8

Observing the Results

. . . children develop ways of finding out what
makes things happen, and what will happen if . . .
(Foster 1983, p. 22)

W e are at another of those crucial junctures where your child will have to perform well if the experiment is going to be a success. Doing an experiment is not as simple as just doing it and getting it over with. Rather, it is about doing it well enough that we are confident we are measuring what we want to measure. You may think that this would have been resolved in the experimental design stage, but in reality, it is not until we actually start to observe and experiment that this becomes apparent. To ensure that we know what it is we're measuring, we'll discuss observations and how they should be undertaken. This is a critical area for record keeping—recording results is not as simple as writing down what you see.

OBSERVATIONS—SEE WHAT WE CAN SEE

To observe and to look are two different things. When we observe something, we are watching it closely; we are attempting to see what is happening with it. When we look at something, we may just glance to see if it has moved or grown or whatever. The degree to which we look at something determines whether we are successfully observing it for experimental purposes. Ensure that your child is actually observing during the course of the experiment rather than just glancing to see if anything happens. This isn't easy, for a number of reasons.

For a scientist, observation means to look at or measure something without affecting the results. This causes problems when we are dealing with things like plants, because just by touching them and stretching them out to measure them, we may be affecting their growth. Great care has to be taken during the observation phase so as not to contaminate the results.

Second, you may find that your child does not quantify things properly. This just means that your child may not yet know how to express changes in number form. Terms such as "bigger" or "longer" or "larger" do not have a numeric value. In science this is crucial. The budding scientist must be able to express how much bigger something is, using acceptable units.

MEASURE WHAT WE CAN SEE

Most scientists use the metric system. If you are from Canada, this does not pose a problem, but in the United States it makes things a little more difficult. The metric system is a system of units that are internationally recognized by the scientific community. You are going to need measuring devices that can measure in metric such as rulers, scales, or volume containers. A short breakdown of the units is shown below, but hopefully your child has been exposed to them before now. These are general conversions, which can be found at the back of most cookbooks and of course your child's science text.

THE METRIC SYSTEM

S.I. Units

(S.I. stands for Système Internationale or International System in English)

Weight	Abbrev.	Length	Abbrev.	Volume	Abbrev.	Equivalent
kilogram	kg	kilometer	km	kiloliter	kl	1,000 of
gram	g	meter	m	liter	l	1 of
centrigram	cg	centimeter	cm	centiliter	cl	100th of
milligram	mg	millimeter	mm	milliliter	ml	1,000th of

To convert to S.I. units, use the following approximate conversions:

Volume: 4 oz = 125 ml   1 cup = 250 ml   1 qt = 1 l
Weight: 1 oz = 30 g   2.2 lbs = 1 kg
Linear Measures: 1 in = 2.5 cm   1 ft = 30.5 cm   1 mi = 1.62 km
Temperature: 32° Fahrenheit (water freezes) = 0° Celsius
     212° Fahrenheit (water boils) = 100° Celsius

We have included the metric system for completeness, but in reality your child should be able to use whatever units he has grown up with. Nonetheless, the use of the metric system will make the measurements more precise (smaller units and divisions), more accurate (closer to the true value), and definitely more scientific. We'll talk later about precision and accuracy in more detail.

We are trying to get the students to *quantify* their results. To quantify something means to give it a numeric value, which can then be recorded and manipulated mathematically to give a better sense of what is going on. Whatever your child is doing—measuring the lengths of plants, the weight of different soil types, or the volume of different gases—all these things can be expressed in a quantity. Even a consumer project will use numbers. Numbers are what separate an experimental project from a nonexperimental one. What we do with the numbers after we get them will determine how well the project will do at the fair.

RECORDING OBSERVATIONS

We keep emphasizing record keeping because it is so important. Your child has to write down all of her observations. Everything she measures, everything she observes, and even things that may have caused problems with her experiment—it all needs to be written down. One important fact that may elude your child is that even if nothing has happened with the experiment, the fact that nothing has happened is an observation. Not only will she have to write down what she sees (or doesn't see), she will also have to record the time of her observation. Keeping a good account of when the observations were made may be important later. Better yet, she should try to do her observing at about the same time each day or within one day if possible. At this stage of recording, getting all the information is the primary job—organizing it in the proper way will come later.

ORGANIZING DATA

All the information that has been gathered must now be put into an acceptable format. The information itself is called data, and in its original form it is sometimes called raw data. Raw data is everything your child has observed happening in his experiment. It is usually recorded haphazardly, listing the time and date, the actual observation (or measurement), and what was happening around the experiment that could have affected the results. Readability and the ability to manipulate this data are the most important points to be addressed here; we'll look at some other issues later. Basically, your child needs to write the data down in some way that will make it more readable, as well as allowing the numbers to be looked at mathematically. A chart will fulfill both these criteria, and a large chart on a wall may give a better idea of what is going on. For long experiments, the data can be transferred maybe once a week, while for short experiments it can all be done at the end of the experiment. This will not be the last time we rewrite this data, so be sure that your child is careful while copying.

ACCURACY AND PRECISION

We discussed repeatability in the last chapter, and now we are going to look at why repeatability is important. Any one particular running of an experiment may contain a number of different errors. Some errors, such as mistakes in reading the instruments or variations due to the weather, can be reduced by repeating an experiment a number of times. Statistically, the more times you repeat an experiment, the higher the possibility that you are actually seeing the answer you should. This assumes that the experiment is properly designed and that all other things are equal as well. Getting closer to the answer that you should get is called accuracy.

A second important concept is precision. Precision deals with how finely we are measuring: how small a division we can read on our ruler or how many decimal places are in our measurement. Accuracy is important for the outcome of the entire experimental process, but precision is important for each and every measurement that your child will take. Whatever measuring device your child uses has small marks on it called divisions.

These divisions represent some unit of measure such as a centimeter or a milliliter or a milligram. The smallest division available determines how precise the instrument is, so if a ruler has a millimeter division, then it is precise to a millimeter. The interesting (and sometimes confusing) thing is that we can estimate a whole level of precision better than that. If the measurement is in between two of the small divisions, we can estimate how far between it is, and get an even more precise measurement. Let's look at an example using the simple diagram of a flower drawn earlier and a representation of a metric ruler.

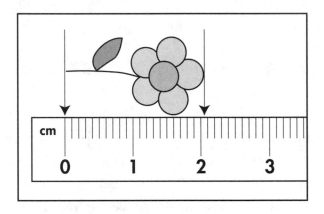

When measuring anything, the proper procedure is to place one end at the zero mark of the ruler and then measure from there. The ruler has its lowest precision at one-tenth of a centimeter, or a millimeter. As you can see, the measurement falls between the 2.0 centimeter mark and the 2.1 centimeter mark. At this point, we can estimate the exact measure of this length and say that it is 2.05 centimeters in length. We have estimated another degree of precision than we had on our ruler. This technique works for all measuring devices that have this sort of linear scale.

TRIAL, TRIAL AGAIN

This title is not meant to suggest that you continue doing your experiment until you get the answer you were expecting, but rather that you may have to adapt your experimental design if you find your answers are way out in left field. When you adapt an experimental design on the fly, as it were, you need to change things around a little bit each time and see if they have a positive effect on the results. This is called the trial-and-error method. Small changes in your child's procedure may be all that is needed to effect the changes that she is looking for, but make sure that she isn't changing too many things or she won't know what fixed the problem. She should also keep good records of these changes and include this problem in her write-ups. Experimental design changes such as these are part of real-life science, and documenting them will enhance your child's project. No judge expects that everything went perfectly from the beginning.

I THOUGHT I MADE AN ERROR, BUT I WAS MISTAKEN

With most things your child is involved in, such as sports or music, a mistake is embarrassing. But in a science fair project, your child should not only recognize errors but report them as well. Keeping track of possible errors is not as easy as it sounds. Some mistakes are obvious, such as measuring incorrectly, or putting the wrong mixture in the wrong pot, or even damaging the apparatus, but many other errors may escape our notice unless we know what to look for. An error in a science fair project is often a variable that we were unable to control. With natural items, the variability could be caused by nature itself. Some seeds will just germinate better than others, maybe due to natural selection (the seed may be of stronger stock) or just because nature is variable (chaos theory).

No matter where the error comes from, it has to be noticed, recorded, and reported. In many cases, scientists do not look for errors until they find an unexpected result—then they search for why it is happening. This is a useful way of dealing with errors for the science fair participant as well. If he finds his results are not what he expected, then he begins his search for sources of error. The most common sources of error are environmental errors, caused by the weather and other such things that are out of your child's control.

EXAMPLES

Let's see how our experimental project is progressing. The frame has been made, and the hair samples have been taken from each person in the family who is participating. Small weights were obtained from the school. Now that everything is in place, it's time to do our experiment.

We've chosen an afternoon to do the experiment and dutifully recorded the time. We carefully attach hair to the frame by tying it around the top. We've tied a weight to a string, and now we attach the string to the hair with the hair

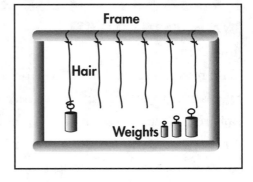

clip. Carefully we let the hair take the weight. If the hair remains intact, we move onto a higher weight, and so on until the hair actually breaks. We repeat this process for the three samples of hair for each family member and carefully record the results.

What did we find? There is not enough difference in the readings because one strand of hair isn't strong enough to hold more than a little weight in all the cases. Disappointment sets in, but discouragement should not, because now it is time to adapt the experiment using trial and error. Think about it, if one strand is not enough, then we should try three or four strands from the same person. We start again; we already have our frame and weights, all we need is the hair. First we decide on how many strands of hair to use, let's say three, then we obtain enough hair to run the experiment more than once. Once again we choose three as our number of repetitions, so we will need nine strands of hair from each family member. Once this is obtained, we can twist the strands together and carry out our experiment again. We will find that the three-stranded hair samples are considerably stronger than the single strands, and that we can see a significant difference between the weight needed to break each sample. This is scientific talk for saying it works.

Now that the experiment is working, we need to check our results to see if they are supporting our hypothesis. Remember our hypothesis was that the older you get, the weaker your hair gets, and we need to see if our results show this or not. In this case, the results sort of support this. Our oldest family member's hair did break before the others, but just barely in one case. Because we did support our hypothesis, we can say that this running of the experiment is a success. Now we have to figure out reasons why one set of our results just barely supported the hypothesis. That's right—it's time to discover the sources of error. We need to take a good look at our apparatus, at our experimental work, at our results, and all the other things going on around us when all this happened, and specifically when we tested the hair from which we got the bad results.

We did the entire experiment in one afternoon, but nonetheless, some of the procedures were done later than others. This could be a source of error since we are dealing with environmental effects, such as heat, humidity, and time of day. These are unintended variables, which we were mostly unable to control. The results were not far enough off for us to consider redesigning our experiment to account for them, so we just mention them in our sources of error. Another thing that could have caused the errors was the amount of time the hair was removed from the head of our subjects. This is another unintended variable, and may or may not have affected our results; we need to remember whether the hair samples we had problems with were fresh or stale. Finally, we need to organize our data into a proper form so that we can see it well and manipulate, if necessary. The form that works most of the time is putting the data in a chart and that's what we'll do here.

If we kept really good notes about our experiment, then we are able to change it as necessary, identify variables we did not control, and basically improve the chances that we are doing a good experiment. This experiment included many of the things we have talked about—repeatability, sources of error, organizing data, and keeping good notes. If your child follows these steps and keeps careful notes, then her experiment should be a success as well.

CASE STUDIES

ERIN

Let's see how Erin is progressing with her trial run of her improved security system, which includes using the existing system's motion detectors and her father's office area, but with the control panel moved to her father's desktop computer. She has everything set up for her trial run; the only question is if it will work. Because Erin has programmed the system to alarm only certain areas, she can conduct her test during business hours, when she can easily get access to the building. Her trial run, which will be repeated several times, involves having a person walk into the alarmed room and timing the program to see how long it takes to perform its various functions. She does the experiment with herself as the perpetrator and her father as the timer. A total of three trials are run and in all three cases the software and hardware work fine. An important thing to note here is that at various earlier stages of her project, Erin has tested her program to see if it is doing what it is supposed to do. It's not surprising that her trial run has worked so well.

• • • • • •

JOSEPH

Joseph has set aside some space in his basement to undertake his experiment. Because of the time factor involved in testing oil-soaked soil samples and the need to prepare the samples some time in advance, Joseph will need this space for some weeks to come. He has started his experiment by preparing several soil samples, each in a separate glass beaker. He collected the soil from different areas around his school and home, and with the help of his uncle, he has identified the different types. He has prepared the oils with which to soak the soil. One oil is like bunker oil found in ships (a common spill) and the other is the type of oil used in cars (a more common spill). Next he soaked each soil sample with the same amount of oil. His brother and father have agreed to help so that they could add the oil to all the soil samples at the same time. Finally, after all this preparation (about a week in all), he adds the enzyme, with his uncle supervising. Every day for the next two weeks he checks his soil samples, weighing each one and visually inspecting them. He has kept meticulous notes and is ready for the next stage. His experiment seems to be successful and probably will not need to be repeated.

• • • • • •

WILL

Will is working on "Does sound affect plant growth?" He obtained some clay pots from his mother, dug some soil out of the flower bed, and got some pea plant seeds from his teacher. He carefully places the pea seeds into wet paper towels in glasses for about a week until they start to sprout. Once this has happened, he transplants all the seeds that sprouted into his clay pots. He neglected to check the length of the sprouts, but at his level this is not a fatal error. Now he sets up his experiment. Just like Joseph he will need a space, one that has access to sunlight, water, and sound. The area around his bedroom window provides the space he needs. He has set up his apparatus on his windowsill and proceeded to insulate certain of the plants from the sound while setting up headphones around the three plants he will use for his tests. The sound dampening is provided by towels and books, but seems to be reasonably effective. He runs his stereo 24 hours a day with continuous radio music, on low volume for one plant, in the middle for the second, and on high for the third. He records his observations every day after school for two weeks, and at the end he has a problem: his non-test plants didn't grow as well as his music listening ones, which is contrary to his research and his hypothesis. He decides to repeat his experiment, using the four plants that weren't tested, with one for each volume and the last one as a control. He has carefully recorded his problems with his initial running and used the same setup for the new experiment as he had for the last. Will didn't think that it was the setup that caused the problem, and his new results prove him mostly right. It works better this time, enough for him to at least be able to report his findings.

EXPECTATIONS

Your child's teacher should have been reasonably involved during the last two stages of the project process. He should have been checking on what your child was doing during the progress-reporting stage and should have checked the procedures planned for the experiment. He should also be available to discuss with your child sources of error, proper organization of data, and what statistics and mathematics can be used to analyze their results. The teacher is an integral part of the project process throughout, and his involvement, although not as important as yours, is crucial.

This phase should have the least amount of parental involvement of any of the stages discussed in this book. For your child to feel like the experiment is her own work, she has to do it herself. Your help should be limited to being a sounding board for problems and ideas and perhaps offering a bit of advice.

Nonexperimental Projects

Science means questioning the world,
wondering how it works, and, while delighting
in its mysteries, raising hope about the possibility
of coming to understanding some of them.
(McNay 1985, p. 18)

N ow let's discuss exactly how to do a nonexperimental project. On our flowchart we're taking the left-hand path. In our opinion, a well-researched and well-set-up nonexperimental project has as much merit as an experimental project. The problem is that many educational researchers disagree. In support of our claim we present this example:

Jerry is a seventh-grade student who is entering the science fair for the first time. Jerry has decided to learn everything he can about Parkinson's disease. His nonexperimental project includes a write-up of

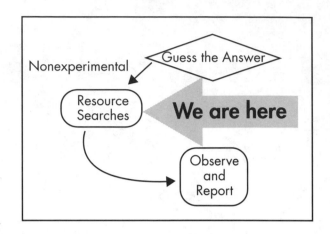

extensive library research and a backboard showing pictures of affected areas of the body and some discussion of the research being undertaken to cure this disease. Oh, and one other thing: Jerry's mother has Parkinson's disease.

Remember that student motivation is a key aspect of a good science fair project. What Jerry has learned while doing this project cannot be understated and has probably been more beneficial than any experimental project he could have done. Jerry's project is just one example, but experiences like his are common.

To aid us in our nonexperimental project, we will use a breakdown of its design just as we have for experimental projects. Not all our steps transfer well into the nonexperimental area so we'll rework them so that they make sense.

STEP ONE: DECIDE WHICH NONEXPERIMENTAL TYPE

Educators with experience in the science fair area can identify many different types of nonexperimental projects. Popular ones include models with explanations about how something works, collections of objects such as bugs or stamps (usually for the younger grades), and studies that involve extensive research to find out a significant amount of information about a topic. Margaret McNay breaks these projects down into the following types:

➢ Presenting three-dimensional displays, reports, and posters based on literature searches.
➢ Building working models or presenting technical demonstrations.
➢ Demonstrating a basic scientific principle.
➢ Observing the environment.
➢ Collecting and analyzing data. (McNay 1985, p. 18)

Your child will have to decide which of these approaches will work the best for his project idea. Deciding what approach to take is just as difficult as deciding which variables to control in an experimental project. Of course, some ideas make this decision very easy, while others will take some thought. In the end, the approach we decide on will determine what we do for the rest of our project.

STEP TWO: DO MORE RESEARCH

Remember those research logs we created and kept (we hope) from when we were determining a question from our idea? Well, now you know one of the reasons we did that in the first place. Being able to go back to original research sources will help reduce the amount of time your child spends at the library. Each book or article he reads may have a bibliography or reference list of books the authors used when writing their book. Finding these other sources is called secondary sourcing. Your child can use this list to find other books on the topic he is researching.

Of course, if your child's approach isn't dealing with a study-type project, then she'll need to research in a different way. If she's building a model, she'll need to find out where to get the materials or how to make the materials. If she's showcasing a collection, she'll need to research what is so special about the collection. If she's creating a working model or demonstrating a scientific principle, then she'll need to know all she can about the original device or principle that she's demonstrating. What she needs to research will determine where she looks to find the information.

So what are some good research sources? Rao suggests: "encyclopaedias, dictionaries, biographical dictionaries, atlases, pamphlets, records, newspaper files, maps, bibliographies, library card catalogues, audio and video recordings, almanacs, textbooks, graphs, brochures, magazines and professional journals, historical stories, photographs and art, charts, magazine indexes, public documents" (Rao 1985, pp. 35–36). These are all legitimate sources.

STEP THREE: FIND MATERIALS

For Jerry's study of Parkinsons disease, to find relevant and current information he'll have to do more than visit the local library. Depending upon the project, writing to a particular association, like the American Medical Association, would be a good idea. If outer space is your child's interest, a good source to write to would be NASA, which has an excellent educational section. If your child wants to build a model, try contacting the manufacturer of the real thing or asking a local engineering firm for specifications. The important thing is that nonexperimental shouldn't mean nonresearch, or nonenlightening. The information is out there, all you have to do is find it.

Another all-encompassing source is the Internet—there are good sites out there. Your child can link to any of the organizations we named above and E-mail a request, for starters. The search engines or searching web pages can help find other organizations you may not have heard about. Some of the information your child needs may actually be available online and this would save a lot of time.

STEP FOUR: WRITE EVERYTHING DOWN

We cannot emphasize enough how important it is to write everything down. Whether your child's project is experimental or nonexperimental, it's important to keep meticulous notes. With nonexperimental research, this becomes even more important, because most of what he is doing is based on someone else's work. If the student quotes someone or uses part of her research or design, he has to give her credit for it. We do that in this book all the time, such as in Margaret McNay's quote at the beginning of this chapter. It is very important that credit be given where credit is due for several reasons: first and foremost because it is the right thing to do. Second, because your child will get caught, and when this happens, he may be disqualified (probably) and be terribly embarrassed (definitely) in front of his peers. Remember that the judges who look at these projects may have read the same research sources your child has, and they'll expect proper bibliographies and quotations. There are several styles, such as APA (American Psychological Association) style or the *Chicago Manual of Style*, which this book uses. Books on these styles can be found in the library or bookstores and will be useful for your child well into his university years.

EXAMPLES

Some examples may help us understand this process a little better, so let's go back to those original questions, but we'll only consider the three we identified as nonexperimental projects: "Where does snow come from?" which converted to "What is the origin of ice crystals in the atmosphere?" or "Why does precipitation change state?"; second, "Why is the ocean blue?" which converted to "How are light and color produced in seawater?"; and finally, "Why are there wrinkles on your face?" which we decided should become "How does the skin age?" Let's take each of these projects in turn through our nonexperimental project process.

Step one involves deciding what type of nonexperimental project we are going to do. For our first question, "What is the origin of ice-crystals in the atmosphere?" we have several choices. To show what happens, we could make a diagram, do a small demonstration of what happens, or a put together a backboard discussion—all equally effective. It depends on what your child is trying to accomplish with this project. If she wants complete understanding, then a written study would probably be best; if she is trying to understand and also teach her classmates, then a diagram is most appropriate; and finally, if she is looking for dramatic value then the demonstration would be best. In reality, all three blended together would make the best project, and that would be our suggested method.

Step two involves more research. By now your child should have a more focused ideas of what he is looking for, basically information that will help him do the type of nonexperimental project he has chosen. In this case, we are looking for diagrams of the process we are studying, information on how to form crystals at home, and written information to support it all. It seems like a lot to do, but we have already done much of the leg work, we've looked through many of these sources already, and we should have a list of sources from our previous steps (it really pays to keep good notes).

Step three is the same whether you are doing an experimental or nonexperimental project: find those materials. What was decided in step one will determine what is needed in step three. You will find this true for many of the steps and procedures: decisions made earlier on determine courses taken later. So we must be careful when we make such decisions that we are not committing ourselves to a course of action that is impossible. In this particular case, all our choices are more than possible. If we are doing a model of clouds, then we need cotton balls or some other cloud substitute. The written aspect does not require materials, but a demonstration requires a number of things. You can make simple salt crystals as a demonstration technique by using salted water, a pail, and a string. As the water dries out, salt crystals will form on the string. This is not a demonstration of ice crystals but rather a demonstration of crystal formation that can be easily explained and demonstrated. Ice crystals are more difficult to form, to keep, and to demonstrate without extensive equipment, so substitution becomes necessary. Always be open to substituting simple materials for hard-to-get materials, otherwise you may be stopped at every turn. Step four is to record everything you did, maybe even with photos.

Our next nonexperimental question is, "How are light and color produced in seawater?" This lends itself to a demonstration, don't you think? Some diagrams and explanations may be necessary but a model or demonstration in this case will have the most impact. Step two isn't very much harder; the amount of research done to come up with the question in the first place was a big help, and don't forget that we did more reading when we were thinking about doing this as an experiment. A diagram of the light spectrum including the wavelength of each color, an explanation of how light diffuses when it enters a medium, and a discussion and a diagram of reflection provide the background for your child's model or demonstration. Now all your child has to do is come up with a small-scale model of an ocean.

Step three is finding the materials for the project. Making a scale model of an ocean will require some specialized materials if it's going to look good. One way of making the model would be to use an aquarium, some sand, small rocks, and, of course, some water. (Your local pet store should have all these things.) You can substitute river, pond, or lake water for seawater, but in reality tap water will do. Then you need some cardboard or bristol board and blue coloring pens or pencils with which to color the sky. Once again, we finish off the process by recording all of our activities—probably backed up with photos and definitely in great detail. When we are recording our process we must be sure to include a justification for the things we are doing and an explanation of why we chose one thing instead of another.

Our final nonexperimental question was "How does the skin age?" Once again, the scope of the question as well as the difficulty in obtaining materials and measurements convinced us to make this project nonexperimental. We also considered performing a consumer project with this idea. If we do not choose this route, then we must continue to consider this topic a nonexperimental one. We need to first determine what type of nonexperimental project we are going to do, and in this case a study (and not just because the word is in our question) seems the best. This study will probably incorporate aspects of some of the other types, including diagrams and photos, for completeness if nothing else. Once we have decided this first step, we start doing more research. We are still looking at the skin's aging process as a whole, so we are now looking for such things as pictures of nonregenerative cells, charts of age data for when this occurs, comparison photos of dried out skin to healthy young skin, diagrams of how the skin folds due to this aging process, and so on. It is easy to see that there are many things we can be looking for while we do our follow-up research on this topic. This makes the next step even easier.

Our research is complete, so now it is time to scrounge. The amount of materials you will need for this project is limited. Basically you are looking for photos, charts, and microscope slides of skin cells (both young cells and old cells, like a before-and-after shot). The cells can either be enlarged to a size that can be seen by the naked eye, or left in their normal state, in which case we would require a microscope. Most of these materials can be photocopied out of textbooks or reference books in the library. For effect, we could cut out pictures from fashion magazines and skin cream advertisements and make a collage. Writing to some of these skin cream companies is not a bad idea either; they are usually willing to share lots of information on these processes. Whatever the case, we have to make detailed notes during the research stage to make sure that we can explain exactly what is going on.

BUT IS IT VALID?

Now that we know how to do nonexperimental projects, we'll close with a discussion of whether or not they are valid for a science project. Not too many years ago, science fair organizers were told outright that students going on to a national level science fair should not be doing nonexperimental projects because they were not of a high enough level for the competition. Part of the reason for this directive was to protect the students from embarrassment. Most science fair organizers view these projects as only useful to get students interested in science fairs, and they think that more should be expected from the better students. To be candid, we once held this viewpoint as well, but after some consideration (and a good look at how some science is undertaken), we decided that nonexperimental projects not only have a place at science fairs but that they are just as valid as experimental ones. Margaret McNay (1985) sees it that way, and her view of science—observing and questioning the world around us—supports her opinion. Eugene Chiapetta and Barbara Foots (1984) agree, giving Einstein and his theories as one example. Who would dare to tell him that his project wasn't of a high enough level? Whether we're testing paper towels, suggesting a new theory of light, or trying to figure out how cancer works, it is all valid and important. "A well researched and inquisitive project addressing any of today's issues can be an informative, problem-solving based, critical thinking enhanced, learning experience without being empirical in nature" (Barron 1997).

The Art of Concluding

*. . . the project comes full circle to the
question that started the adventure.*
(Hamrick and Harty 1983, p. 46)

The experiment is done, the results are in, and the winner is . . . well we don't know yet because we haven't looked at our results to answer the questions we asked. We are nearing the end of our science fair flowchart and we have to start answering some questions. But just calling them answers is not correct; these are conclusions.

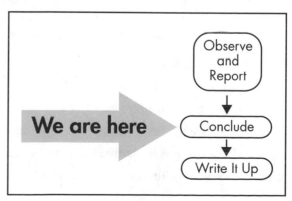

The process of concluding has its roots in the inductive reasoning process, where concluding is linked to inference. An inference is just an interpretation of the results produced by the experiment that your child has undertaken (Oswald and Preyra 1990), while inductive reasoning is one of the philosophical theories on how people use their minds to solve problems and answer questions. Basically said, you have to look at what came out of the experiment, your observations, and decide exactly what it means. This may seem simple, but as in all things scientific, appearances may be deceiving.

Once again we will break this process into several steps and take them one at a time. These steps are comparing the results to the hypothesis, stating what is actually seen, understanding the changing nature of science, asking new questions from the old, and nonexperimental approaches to conclusion writing.

TAKE ONE STEP FORWARD, BUT LOOK TWO STEPS BACK

Before we can begin to state our conclusions, we have to look again at our question and the answer we proposed for it. This is called reviewing our hypothesis, and it involves comparing what has actually happened to what was expected to happen. If we make a prediction in science, we always have to back up that prediction somehow. In our case, we are backing it up with further research or an experiment. So we need to find out if our results supported our answer, and if they did, we have to state that clearly in our conclusion. If they didn't, we need to say that as well and suggest some reasons why not. Don't forget that an experiment that failed to support its hypothesis is sometimes as valuable as one that succeeded. In either case, it can lead to further research, which we'll discuss later in this chapter.

WHAT HAVE WE SEEN?

Many students in the science fair will start their conclusions like this, "We expected to see . . . but were not able to because . . ." Although this bases the conclusions on the hypothesis, it comments more on what result the student wanted to see rather than what he did see. Your child has to base his conclusions on what he has seen, not what he expected to see. If he has an unexpected result (which scientists call a discrepant event), he needs to comment on what he actually saw in that discrepant event—not what he expected or wanted to see. Only after he has commented on his actual results should he mention that it was unexpected and did not agree with his hypothesis.

One of the major reasons that students should comment on what they actually saw—rather than what they wanted to see—is to assure the judges that the experiment was not tainted by the student's preconceptions. Preconceptions can cause even experienced researchers to see something that didn't actually happen, because they expected to see it. Any experiment that involves human observation can fall victim to this experimental error, and in some cases this taints millions of dollars and thousands of hours worth of important research. This is not to say that your child still will not skew the results based on what she wants to see, but she can't proclaim it obviously for the judges and her peers to see.

SCIENCE ISN'T BLACK AND WHITE

Even if our results give us a conclusion in line with our hypothesis, this only supports the hypothesis—it does not prove it to be true all the time. Science does not answer questions definitively, but rather supports theories through experimentation. Nothing is actually ever proven in science, only accepted for the time being until new information makes the theory obsolete. Susan Bosak, author of the popular science activity book *Science Is . . . ,* explains this very well: ". . . science involves much more than dispensing scientific 'fact.' In fact, very little in this world is 'fact,' particularly in science! The more we learn the more 'fact' changes to reflect our new knowledge and perspective on reality." (Bosak 1991, p. 12)

An excellent example of this is the work of Aristotle. Many consider Aristotle the first scientist—he looked at the world around him and attempted to explain what was going on in terms that many could understand. This role of a scientist has not changed, but the theories of Aristotle are no longer accepted. Aristotle believed, based on his own experiments, that the substance of the world was made up of four elements: earth, air, fire, and water. He also believed that the reason things fell to the earth was that they were returning to their natural place, and that the larger the object the faster it would return to that place. These theories were accepted as fact for more than a thousand years, but when contrary evidence was produced, they ceased to be fact and became yesterday's news. This illustrates something about science that few really understand—that science is not totally made up of facts, but is also made up of theories supported by evidence that may be discounted at any time. The scientific community accepts these theories (for the most part) until they are proved false. The same needs to be said about your child's hypothesis, he hasn't proved the hypothesis, he's just supporting it for the time being.

NOT ONLY ANSWERS, BUT QUESTIONS TOO

Sometimes when we do an experiment, we open up a Pandora's box of questions. In fact, many experiments will wind up asking more questions than they were designed to answer. This happens because as students (and parents) do experiments, they see other things happening than what they expected. These can lead to further research questions for future experiments, but they also form an important part of the conclusive statements. Asking questions about results and experiments as a whole shows that students truly understand what they've done, and that they are thinking beyond it.

CONCLUDING NONEXPERIMENTS

Conclusions take on many forms. For experimental projects, the conclusions are a direct result of the observations made during the experiment and the hypothesis suggested earlier. Nonexperimental projects end up concluding what the student has learned through the process.

We could set up a nonexperimental project just like an experimental one with a hypothesis and all, but the fact is, it's not as appropriate. The nonexperimental project takes on a purpose rather than a hypothesis, and the purpose of a nonexperimental project is usually learning about something. So when we conclude in our nonexperimental project, we should be discussing what we learned. Did any new facts come to light that we weren't aware of? How does recent research in this area compare with earlier research? How does this topic impact society? We discussed asking questions and proposing answers to nonexperimental projects during our experimental design process, and these can be useful for making conclusions as well, but in the long run stating a purpose for the nonexperimental research is more effective and will make the conclusive process a little easier.

JUSTIFYING THE CONCLUSION

The form that conclusions should take is almost as important as the conclusions themselves, because if they are not written in the proper form, they are more difficult to justify and also to read. The well-written conclusion will not only make the statement, but justify it as well. Each statement must be based on results or observations, and justified by them. Unsupported conclusive statements are big no-no's in the science world, and because a science fair emulates that world, they're big no-no's in the science fair world as well. Judges will be quite harsh on students who have not justified their conclusions properly, whether in writing or during their verbal presentations. Think of it as being like accusing someone of a crime with no evidence. A court judge wouldn't be happy with a police officer who did this, but checks and balances in the legal system try to prevent this. In a science fair, there are only two checks in this process: you and your child's teacher. The teacher has to oversee a class full of projects and is more concerned with making sure the students did them at all and in the proper form than if they have supported their conclusions, so the matter lies with you.

CONCLUDING OUR EXAMPLES

We are now at the last step for our experiment on the tensile strength of hair. You will remember that the experiment went all right, and we saw much what we expected to see in our results. This should mean that our conclusions will be straightforward and easy to write. Also remember that we repeated our experiment and checked our results to see if they supported our hypothesis before finishing up. This is one method of helping with conclusions and making an experiment seem to go smoother, but in truth not all experiments can be adapted to support the hypothesis. In fact, our experiment just barely agreed with our hypothesis, and this has to be mentioned in our conclusions. We don't have to worry about commenting on what we wanted to see versus what we did see because they were one and the same. We can easily write our conclusive statements based on the hypothesis and the question we asked in the first place, and our results support this as well. An example of the conclusive statements may look like this: "We found that the older the donor of the hair, the less weight it could support. We also found that a single strand of hair could not support enough weight to do the experiment. These results support our hypothesis and allow us to say that, at least in our family, age makes the hair grow weaker."

This statement does not go too far, it does not make any generalizations that it cannot support, it justifies its conclusions, and it addresses the hypothesis—just what we are looking for. More can be added of course, if it is necessary. No questions were asked in this conclusion, even though many manifested themselves throughout the process. We studied age as a variable but we did not consider the sex of the donor, nor did we look outside our own family for donors. If we want to expand on this experiment, those are two ways we can do it, and your child could enter much the same project next year with this expansion. Speaking of next year is a bit premature, but keep in mind that a project that can span more than one year is a project that will be stronger in each successive year, with less effort as well.

We have made our conclusions for our one experimental example, but now we have to look at our nonexperimental projects as well. Our nonexperimental projects were:

Where does snow come from?   What is the origin of ice crystals in the atmosphere?

| *Why is the ocean blue?* | How are light and color produced in seawater? |
| *Why are there wrinkles on your face?* | How does the skin age? |

In each of these cases, we established a question to answer that was the purpose of our research, and because we originally looked at all of these ideas as possible experimental projects, we also suggested hypotheses. For our purposes, the hypotheses are not necessary, for in a nonexperimental project we do not need to speak to the hypothesis but rather state what we have learned in the process.

For our first nonexperimental project idea, "What is the origin of ice crystals in the atmosphere?" we would write a paragraph on how precipitation is formed in clouds, and how that precipitation falls to the earth. We would probably include a write-up on the states of water and the whole water cycle with our project, but it is not necessary in the conclusions. The art of concluding in a nonexperimental project is not in what to include but in what not to include. In this case, we should stick to the facts: water vapor changes state in the atmosphere forming water droplets and clouds. When the drops get to a certain weight they fall. During their fall, or before it, they may freeze. In either case, during the journey to the ground they change state once again into a solid crystalline form we know as the snowflake. The conclusion would include everything you see written in explanation above, because this is what we learned from this project.

Our next nonexperimental question is: "How are light and color produced in seawater?" Once again we have a hypothesis. Although in the last project we said that a hypothesis wasn't necessary to tell what we learned, in this case the hypothesis can play a role, because in many ways this is a very "experimental" nonexperimental project. This means that the explanation of what is happening with light in seawater can easily, and probably more efficiently, be carried out by using the experimental method even though this is a nonexperimental project. What is happening with light and seawater is that the sky is reflecting its color on the surface of the ocean, much like a mirror would reflect color. This is why on stormy days the ocean seems gray rather than blue, and our little demonstration model supports all of this. So our conclusions would look something like this: "Observational evidence supports the hypothesis that the color of the ocean is a reflection of the color of the sky. A demonstration to this effect has been undertaken and also supports this hypothesis."

Our last project had the possibility of being a consumer project, so we are going to look at it from several different angles. First we will deal with the nonexperimental side of the question: "How does the skin age?" We have learned that as the skin ages, its ability to replicate cells has diminished, and that the skin cell membrane, which surrounds the skin cell itself, has weakened and is drying out considerably. We also found out that as the skin ages it thins considerably, and that many of the skin functions that diminish with age are hurried along by exposure to sunlight. There are many theories about treatments and actions we can take to aid our skin in its fight against aging, such as protecting the skin from the sun as much as possible and using vitamin E cream. Finally, we could get a face-lift, which actually stretches the skin of the face to "smooth" out wrinkles. For this nonexperiment our conclusion would be: "Skin will age no matter what we do. During this aging, the membranes around the skin cells dry out and, much like an apple left too long in the sun, wrinkle up. The thinning of the skin and the excess skin that then becomes available to the face muscles causes the wrinkles. Theoretical preventive measures include protection from the sun and the topical use of vitamin E, but as a last resort plastic surgery could be considered."

This is a very complete conclusion, commenting on what we have learned as well as offering advice. We'll now look at our possible consumer project, which was testing various types of soap for their cleansing and moisturizing abilities. Here our conclusion could include all the information given above (done as background research) as well as which soap won out and why. That statement would be fairly simple: "Brand X was the most effective shown by our tests."

CASE STUDIES

ERIN

Erin has many good things to report, since her system test worked perfectly—it phoned the police dispatcher and the manager's home when an intruder entered the part of the building where the alarm was armed. Erin was careful to record the reaction time of her system and compare it to the human-operated one. The existing system is quite fast, with an average response time of two minutes. Erin's system was much faster: it phoned the police an average of thirty-five seconds after entry of the intruder, and the manager forty-one seconds after that. But it isn't perfect—the police and the manager point out that there is no recourse for false alarms, while in the existing system a false alarm can be canceled with a phone call to the alarm company. Also, the computer she needs for this system is quite high tech and even a basic level operator would require some training on it. The final problem with the system was pointed out to her by her brother: what if the thief takes the computer? These are questions that Erin can use for further research and include in her conclusions. What should those conclusions look like? "The computer-operated alarm system performed faster than the existing system, while using some of that system's hardware. Although this supports our hypothesis, some questions were raised about learning the new system, false alarms, and possible theft of the computer. These are all valid points, which we believe can be addressed to make this system more economical yet more efficient."

• • • • • •

JOSEPH

Joseph has not been idle either; he has been carefully recording everything that has happened with his experiment every day for the past two weeks. Checking the soil samples, weighing each one, and observing the changes occurring has led him to believe that the experiment was successful and will not need to be repeated. His observations show that the enzyme was successful in breaking down the oil to some degree in all the cases, but there were some differences associated with the different soil types and the different types of oils. This is just what he wanted to hear. This means that his conclusions support his hypothesis, and that there is evidence that the enzyme will work with all soil and oil types, just at different rates. Because this experiment has such far-reaching goals and is fairly cutting edge, Joseph takes some time and care with the writing of his conclusions. "The enzyme used in soil remediation worked for all the soil and oil types tested, showing beyond the shadow of a doubt the economic and environmental importance of this new chemical. We observed that although the enzyme had a noticeable effect on the oil in all cases, some of the soil types slowed the rate of breakdown. Sandy soil proved to be the easiest type for the enzyme to work in, and we would suggest that this type of soil be used primarily around areas that may experience oil spills."

Once the conclusions were completed, Joseph showed them to his uncle who helped with some wording and removed the words "beyond the shadow of a doubt." His uncle reminded Joseph that no experiment—no matter how well undertaken—"proves" anything; it just supports or disagrees with a hypothesis.

• • • • • •

WILL

Will, you'll recall, had some problems with his experiment: "Does sound affect plant growth?" His control plants did not grow as well as his test ones so he repeated his experiment. His findings the second time around were better, but not amazingly so. Will is now left with only a barely supported hypothesis and the knowledge (from his mother) that he probably hasn't given his project enough time.

This is a common problem for many science fair participants: they do not leave enough time for their projects to run properly, and they especially do not leave enough time for a repeat of their projects if necessary. That is why we included a timeline earlier in this book, to give you and your child a better sense of the timing for a science fair project. Even though his project is weak, Will has to go on.

Will has decided to comment directly on his hypothesis. He has also come to the hard decision of criticizing himself in his conclusion. Your child's own faults (everyone has them) are sometimes an important source of error. They manifest themselves in statements such as, "I didn't water the plants each day at the same time," or "I didn't leave enough time for the experiment to run properly." Will's conclusions will end up looking something like this: "Our hypothesis stated that sound, particularly the volume of sound, would have an effect on the growth of plants, and this was supported by our data. Three of the plants, the control and the two that experienced the least volume of sound were close to the same length of growth after two weeks, but the plant exposed to the most sound volume was shorter by 1.5 cm. We realize that the time frame was too short for this experiment to run properly, but we had to repeat the process after a failed first attempt. We believe that the setup and apparatus used were fine though, and can be used to repeat this experiment at a later date.

We will visit our three budding scientists later as they put the finishing touches on their projects. They are not finished yet, even though their experiments are completed and their conclusions written. Now they have to deal with their report writing, backboard preparation, and their own preparation for their respective school fairs.

CONCLUSIONS AND PARENTS

Your child has done her experiment and now she is turning to you for help with her conclusions. Be careful! Even here, too much of a good thing is a bad thing. As a parent you have to assist the teacher, who is unable to give your child individual attention at this stage, but you have to do this with wisdom and restraint. The easiest thing for you to do is tell your child what to write, but that is also the worst thing you can do. You've worked hard not to interfere up to this point. If you do too much now, you'll be sending a message to your child that she will never have to truly complete anything because you will always do it for her. That may sound a bit far-fetched, but it happens far too often. You can help your child at this point by asking pointed questions that will bring out her own conclusions. Good questions are: "What did you see?" "What did you expect to see?" "What happened in your experiment?" "Why did/didn't your experiment work?" or just plain, "Why?" As you discuss these questions, make sure she can justify her statements with evidence from her experiment or research. In this way, you are a catalyst for your child's learning, and as she learns more, so will you.

As a parent you cannot make the conclusions for your child, but you can ensure that the conclusions are justified and that they make sense. You can also help with wording of the conclusion. Finally, you can remind your child what he should be considering when he writes his conclusion. Basically, you can help with everything but actually drawing the conclusion.

Writing the Report

Your report should contain all the important aspects of your science project.
(Connors et al. 1988, p. 40)

Entire texts have been written on just the topic of writing a scientific research paper, so we will only touch on some of the points needed to create a good one. The scientific report that accompanies a science fair project has some differences from one that stands on its own. For starters, the science fair report will have visual aids accompanying it (the backboard) and will provide a chance for the students to defend their work. The report in this case does not have to stand alone, which is rare for other types of research projects, unless some sort of public speaking is involved. That being said, some teachers prefer to grade a stand-alone report rather than grade the students' science fair projects as a whole. Dissuade them of this if possible—a science fair project as a whole is a major undertaking and should not be dismissed for grading purposes. The whole body of work, not just the written report, should be included. Teachers may attempt to sidestep the issue by making the science fair aspect of the project voluntary, but once again the whole body of work should be considered, not just one part in isolation.

The report itself serves a number of purposes. The report is a summary of all the work that your child has done. It incorporates everything—from the first day to the last—in one document. The second purpose is to remind your child of every aspect of the project. She's been working on the project for many months and may start forgetting important things about what is going on; writing the report will ensure that won't happen. Finally, the report is something your child can keep from this year's project, while most everything else will end up being returned to its proper place. In the end, when the materials are gone and the backboard is stripped, the report remains as a record. It may even serve as a starting point for the following year's project.

DON'T QUOTE ME!

Students should not copy statements directly out of their reference sources unless they are properly put in quotes. This is a common problem for teachers and students in the school system. Kevin Collins, a teacher in Idaho, used to find that many of his students when asked to do reports, ". . . copied material straight from the book, often not even reading over what they had copied" (Collins 1981). Students need the skill of paraphrasing. Paraphrasing is the ability of a student to read a section or paragraph from an article or book, and put it into his own words. This skill should not to be taken lightly. Try it yourself; it is difficult to not write out exactly what you have read. What the student has to do is read the material over several times until he has an understanding of what is said, then say out loud what the section means. He can tell this to a family member or just talk to himself, it doesn't matter. As he becomes more practiced, he'll be able to do it without speaking out loud. Then all he has to do is write down what he just said. Nearly all the time, what he writes will be in his own words, yet will retain the original meaning. It still needs to be referenced (which we'll deal with in the next section), but now that he understands what has been written, there is no need to quote directly. You may think that this is an awful lot to be going through for this part of the project, but the skill of paraphrasing is important to your child's entire academic career; he will use it until he graduates and beyond. This is one of many important skills that your child can gain by doing a science fair project.

CREDIT WHERE CREDIT IS DUE

When we were doing our initial research, we kept good notes on where we found our information. That will pay off in this section. Every bit of information that your child has used from a published source (books, magazines, newspapers, etc.) must be properly referenced and quoted if necessary. If it isn't, then there may be problems. These problems stem from a common one in the school system: dealing with plagiarism. Plagiarism is what happens when students take the work of others and call it their own. It is a serious offense, at higher grades it leads to failing grades, suspensions, and—beyond high school—even to expulsion. Your child can use the material she researched as long as she gives credit to the actual author through referencing. We discussed this in chapter 9, but we need to emphasize how important this idea is now that we have to actually start referencing. Throughout this book authors are quoted, and they are properly referenced. Some authors just contributed ideas, and they are also included in the references and bibliography. No matter what type of project the student is doing, references to sources need to be included in their report.

Another type of credit should be given as well. If your child was lucky enough to have the mentorship of scientists or engineers for his project, he should give credit to them also. This can be as simple as including their names in an acknowledgment at the beginning of the report or the backboard, but it should be done.

THE PROCESS

Once again we can use the process we established since the beginning of this book to help us do our work. If we are doing an experimental project, then we can write up our research paper in much the same way as we have done it. For style purposes, we can do this for nonexperimental projects as well, as long as we remember that the amount of actual research information required will be greater for the nonexperimental project. The "scientific method" that we laid out in chapter 3 is listed below. We have omitted the write-up and reporting stage and the backboard preparation stage for the obvious reason that each of these sections will come together to form a section of the formal written report that is separate from the backboard itself. They also correspond to particular chapters in this book and stages of work that the student has been doing.

As you can see by looking at them side by side, there are many similarities, in fact the first three sections and the last two sections are exactly the same. We'll focus on these similar sections first.

THE EXPERIMENTAL PROCESS	THE NONEXPERIMENTAL PROCESS
Idea generation phase — come up with the idea for the project	**Idea generation phase** — come up with the idea for the project
Idea research phase — find out what you can about your idea to help with the next steps	**Idea research phase** — find out what you can about your idea to help with the next steps
Question formulation — turn the idea into a question that can be answered	**Question formulation** — turn the idea into a question that can be answered
Experimental design — design an experiment to answer the question	**Resource searches** — find out as much information on the topic as possible
Experimental activity — do the experiment	**Experimental activity** — N/A
Observations/results — watch and record what happens	**Observations/results** — watch and record what happens
Conclusions — the answer to your question	**Conclusions** — the answer to your question

IDEA GENERATION

This section provides a very nice way to begin the research paper. It allows the student to comment on where she got the idea for the project and why she considered doing this project in the first place. This gives the project its purpose and origin, which could be the heading for this section rather than idea generation. The headings for each section have to have meaning, but are not as important within the written report as they are on the backboard. We will be looking at this in the next chapter, but as long as everything is organized in a scientific-method sort of way it should be all right. There should not be much detail about the actual project, because that will be covered in the next section. But your child should talk about what she had to do to come up with the idea, who helped her come up with it, and why she decided on this idea in the end. This will also be good preparation for the interview process she'll undergo at the fair.

IDEA RESEARCH

Some call this background research, which is another possible title for this section, but in the end the student is researching his idea to find out what aspect of it he will do his project on, so idea research is more appropriate. In this section, your child should start writing what he found out about his idea. Nothing is too general here, all the information he collected is valid, but remember this is just a prelude to the main work, so it should be relatively short and concise. Listing facts he has learned is one effective way to prepare for this section. Because he kept good notes (right?), he should be able to figure out what he learned about the idea before the main research and exactly where he got it from (research log). These small tidbits of information are what belong here.

QUESTION FORMULATION

This section brings together our first two areas to form the basis of why we did our experiment—essentially the purpose of the experiment. It is useful to recognize these common word forms of the scientific method for the next chapter on backboard, so we will continue to point them out to you. In this section, the student not only made up her question, but proposed an answer to it as well. In her report she will have to explain why the idea and the background research made her decide on this research question. She can bring in ideas about the scope of the project, why she decided to do an experiment or not, and exactly how the question came into being. Once this is accomplished she has to write about her hypothesis.

Choosing an hypothesis wasn't easy in the first place, but having to write about it might be even more difficult. Some students have trouble putting into words what made them choose the specific answer they did; it sort of comes as a feeling after their background research and may be difficult to justify. But justify it they must. The written report must contain some justification of the hypothesis, and that justification has to come from the research, not from some feeling, or opinion. Remember, the hypothesis is an educated guess, with more emphasis on educated, and less emphasis on guess.

Our research road forks at this point because our sections no longer match for the experiment and the nonexperiment. We'll deal with the experimental write-up sections first and then move onto the nonexperimental ones.

EXPERIMENTAL DESIGN

Now to the meat of the matter: describing how we designed our experiment. To properly write up this section, it would be a good idea to refer to the chapter on experimental design. We set down a process in that chapter that could help the student write out his own design process. Let's review our four steps of experimental design. Step one: identify what will be tested; step two: identify what will not be tested or changed; step three: identify the materials needed and obtain them; step four: write it all down. These four steps can easily be used to break down how the student designed the experiment.

Of course we cannot use the plain language of this book, instead we must use scientific language. So rather than identifying what will and won't be tested, we'll identify all the variables in the experiment, explaining which will be controlled and which are varied. We can still use the word "materials" though, or "apparatus," whichever you and your child are comfortable with. But we won't be writing everything down, we will be recording our design process. We can now refer back to that experimental design sheet we used (we hope), because it has most of the information we are looking for on it already: our variables, controlled and otherwise, and our materials. Listing the materials used and their sources is an important part of this section of our report, since this gives the judges and the teacher some insight as to why we got the results we did. It also shows how imaginative and creative students can be when they have to scrounge for stuff. As these are the actual steps your child may have used to design her experiment, they should be a good resource when she writes about designing her experiment.

EXPERIMENTAL ACTIVITY

Now that we have explained how we designed our experiment, we have to explain in minute detail what we did with that design. This is the section that deals with running the experiment: everything the student did while setting up the experiment, carrying out the experiment, and finishing up the experiment. It is commonly called the procedure, and it should be fairly long but still concise. Once again, a step-by-step or blow-by-blow account of what was done is the best way to go. Your child should talk about how carefully he kept his controlled variables controlled, and how he took his measurements. The tools used, measuring devices, time of observations, and just about everything you can think of about doing the experiment should be included. No other section—except the observation section, which may include some raw data—should have as much information in it as the student's own summary of the work he has done. That's right, this is a big part of telling the judges and the teacher how much planning and work went into this project. Earlier we mentioned making diagrams and taking pictures of the experimental apparatus to ensure that it could be set up again if needed, and also to give a better sense of what the setup looked like. Pictures of the student's actual experimental setup can be included in the written report here, especially if the apparatus was too large or fragile to be transported to the science fair site. This is one place those pictures will come in handy, the other being on the backboard itself.

OBSERVATIONS/RESULTS

We are getting closer to the elusive end of the road for this part of our project. The observations/results section is the next to the last section that has to be written. Here we can do a number of things. We can display our data in a variety of formats: as tables, graphs, mathematically manipulated numbers, before and after pictures, or whatever. An experienced group of science fair organizers broke the possibilities of written results into three broad categories: written descriptions, graphs, and charts or tables (Connors et al. 1988). It all comes down to what is most appropriate to the data. A written description of what is seen is an excellent way to record observations of nature, or even some types of experiments. Experiments involving nature, the world around us, our environment, or even something like a color change would probably be best served by a written description. This would entail the student simply writing down everything seen. This is raw data, but it's very appropriate in some cases. Raw data should be included only if it is short and sweet; multiple pages of raw data should be put into an appendix instead. The data being shown should have been analyzed, and one surefire way of looking at data is by using graphs. Each type of graph has a different purpose. Below are a few examples.

Line Graphs are usually used to compare two things that are related to one another mathematically. An example would be our tensile strength of hair experiment, where we compared the age of the subject to the amount of weight their hair can support. An actual line graph from example data is shown below.

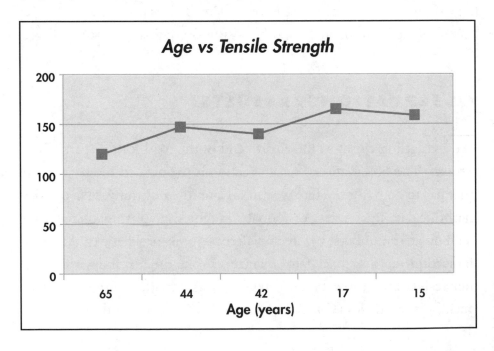

Pie Graphs are used to compare several parts of a whole, as in pieces of a pie. The whole pie is 100 percent, but each individual slice of the pie is a different percentage. Any data that involves percentages would work well in a pie graph. A good use for such a graph would be comparing the results of a classroom survey. An example of a pie graph used for a survey is shown below.

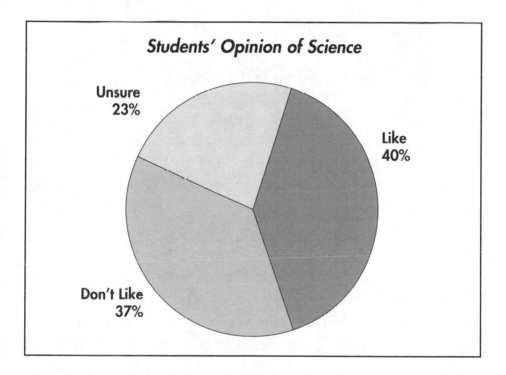

Bar Graphs, like line graphs, are used to compare two related quantities, but in the case of a bar graph, there does not have to be a mathematical relationship. The bar graph would be appropriate for Will's experiment involving the length of different plants on different days or for tracking something over a number of days, like the movement of birds south for the winter. An example of a bar graph used to track birds flying south is shown on the next page.

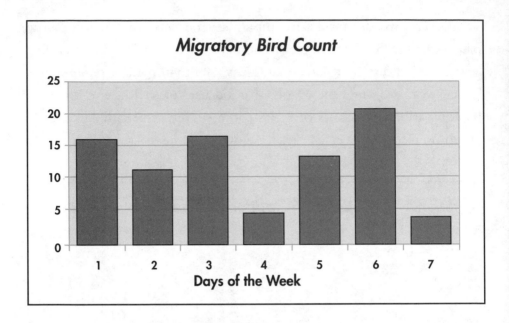

The neat thing about graphs is that if your child is keeping her numeric results on a computer in a spreadsheet or database, the program can make the graph for her. Even doing graphs by hand is not very difficult. This is an important math skill usually learned by junior high.

The final way students have to display their data is in charts and tables. Although this may seem low tech or less exciting than the graphs above, even the data for those graphs needed to be stored in a table. Just because a graph is used doesn't mean that the entire data table should not be present as well, as some judges and teachers want to see the data in its numeric form. Data charts and tables are relatively simple; just organize the data much the same way it was collected. Most data is time dependent, meaning that when the measurement was taken is important, so just list the data next to the time it was taken. Put a heading for each column on top, outline the whole thing if necessary, and that's it.

The types of data that involve direct student observations can be portrayed with drawings or photographs. Once again, a picture is worth a thousand words, and let me tell you, it's easier to include a picture than to write a thousand words!

CONCLUSIONS

We've spent an entire chapter on conclusions so we'll be brief here. Basically, the statements themselves are written here with their supporting evidence. The evidence can include the results and observations, the graphs used to analyze the data, the raw data, the experimental procedure, or any of the diagrams and pictures that have been taken. This is the evidence that will support the statements in the conclusion. The other thing that has to be incorporated in this section of the report is sources of error. We have discussed sources of error in several chapters, and we have commented on the need to keep track of everything, good or bad, that happens in an experiment. Now is the time to put the bad stuff in writing. Sources of error can be blamed for just about anything: strange results, bad results, unsupported hypothesis, or just the fact that an experiment didn't work. These sources are many and varied, and they fall into several different categories: observer error, experimental error, environmental error, and natural variability.

Observer error, as the name suggests, involves all those errors that occur due to the fact that the experiment is being watched by a human being, and all human beings and their observations are flawed. These errors might include misreporting or misreading measurements, not measuring the right thing at the right time, variations in measurements, and so on. Anything that the observer has done is subject to suspicion, especially if he is observing something in nature, because just by being present he affects the results.

Experimental error is a little more difficult to identify, because it involves mistakes made in the design or setup of the experiment. Some variables might not have been identified or properly controlled, some variables might have been tested that should not have been, and any of these can cause problems with the results.

Environmental errors involve errors and variables that we cannot, for the most part, control. As the name suggests, these include the weather, air pressure, humidity, temperature, noise, air quality, all things dealing with the environment where the experiment was held. Some can be controlled with much effort and money, but for the most part they remain overlooked or just deemed unimportant, until of course the experiment doesn't work the way it should have.

The final source of error is natural variations. Any experiment will yield slightly different results every time it is performed, and these are usually due to natural variations. These variations do not play a big role in the experiment, because the results are usually large enough to see the difference in spite of the small variation. In either case, it is a staple source of error used in many science fair projects, and its inclusion in the project write-up won't hurt.

Now we'll discuss the nonexperimental report writing section, continuing from where we forked off into the experimental area. This discussion starts with our resource searches and ends with our conclusions just as in the experimental project.

RESOURCE SEARCHES

Here students can talk about where they looked for their information, how current it was, and what type of information was gathered. Showing initiative and imagination in finding resources is rewarded in nonexperimental science fair projects, so it should be documented. Students can talk about interesting Internet sites, or the public library they used, or the scientist or organization they contacted, or just about anything they used or did to get the information they used for their project. This should not be too difficult if the student has kept good notes.

OBSERVATIONS/RESULTS

The observation and results section of the nonexperimental project contains the information the student has collected that is relevant to his topic and to the answer to his question. Don't include irrelevant information. Information pertaining to the history of the research or the subject can be a good lead-in, but it doesn't belong in the observations/results section. It would be better in something like a "Body of Research" section. Remember that all the research that has been taken from other sources has to be properly referenced and a bibliography has to be prepared. The lack of experimental results is what some feel is the fundamental stumbling block for the nonexperimental project, but in this report-writing process we deal with this lack as well as can be expected.

CONCLUSIONS

We have reached the end point of our nonexperimental project and must state our conclusions. Once again this has been covered in some detail in the last chapter, but there are several things to remember. These conclusions must be based on the body of research that was presented in the previous section. Opinions have no place here. Each conclusive statement has to include the evidence that supports it, and these statements have to be concise and to the point. Although there are no sources of error in this type of project, there might be dissenting research or viewpoints. Some scientific works have been argued about since they were published. Einstein's theories are one example, for at the time of their publication, many scientists did not agree with his work. Earlier we used the example of cold fusion; there are many dissenters to this theory and the experiments that support it, but many people agree with it as well. Sometimes there may be research that disagrees with the research you used to support your answer, and if such exists, mentioning it in the conclusion is a nice way of showing how tentative science is and how complete your research was.

The headings suggested within this chapter are not written in stone. These headings are just meant as guidelines to show what sort of breakdown can be done. The written report is a very formal type of writing and should not be undertaken lightly. Many educational researchers suggest that some practice reports be done over the year (Bombaugh 1987). Most high-school teachers will have done this in one of their science courses' labs. Many senior high (tenth through twelfth grade) science courses require that a lab be written out in a formal report manner, but the junior high (seventh through ninth grade) student won't have this experience until the higher grades. Suggesting that this type of formal lab report writing be incorporated into the curriculum is one possibility. The student's teacher makes all the difference in these cases. Parents can take an active part in their child's education by making suggestions to the teacher.

Creating the Backboard

*Once the project is completed, the next step
is to set up a display to show the course of
the investigation and explain its outcome.*
(Hamrick and Harty 1983, p. 47)

The report is finished, the project nears completion, and all that remains of the preparation is the creation of a backboard. What is a backboard? It's a display area that students use to show their projects to everyone who walks by the table. As the name suggests, it is a freestanding poster board, usually divided into three segments that resides at the back of the table. A few people may stop and read a report, but many will take the time to look at posters lovingly glued onto plywood proclaiming the project for all to see. Creating the backboard falls within the reporting stage of the project, just like the written report. In fact, it has many similarities, including organizing information and presenting information for review. It might help to think of it as a shorter, larger version of the written report.

Creating a backboard is not as simple as it may sound, as there's a lot of information and only so much space. The good news is that there is a method to this madness as well: creating a backboard will break down into the student's description of the scientific method. The process we have laid out in this book will form the crux of a good description of a scientific method. Let's review this process to refresh our memory:

> *Question formulation:* turn the idea into a question that can be answered.
>
> *Experimental design:* design an experiment to answer the question.
>
> *Experimental activity:* do the experiment.
>
> *Observations/results:* watch and record what happens.
>
> *Conclusions:* answer your question.
>
> *Write-up/reporting:* explain what you did.

This is quite similar to the process we went through to write our reports in the last chapter. But the backboard is a different sort of animal—it has to summarize the student's work at a glance. It must be dramatic and eye-catching, it must contain enough information to tell the story but not so much that the reader gets lost, and it has to do it all in a space smaller than a tabloid newspaper.

A closer look at each of our steps is necessary to get a better handle on what we need to write.

Question Formulation

We devised a question to answer and proposed an answer to it. These have to be stated clearly so that everyone can see the purpose of our work.

Experimental Design

This is what we did for an experiment; the procedure and materials we used for the experiment.

Experimental Activity

This is where we finally got around to actually doing the experiment; it would explain the procedure in more detail, and comment on false starts.

Observations/Results

This is our results and observations. Here we explain what happened in the experiment, what we saw, and what went right and what went wrong.

Conclusions

This is the point where we compare our results and observations to our purpose and our hypothesis (the answer to the question we posed originally). Here we state our conclusions based on our results and observations and the sources of error that we encountered.

The bolded words are your backboard headings. These headings are commonly used on thousands of backboards in hundreds of fairs every year. Each heading is part of the process that by now should be familiar to you. We will use each of these headings for our descriptions of what and how much to write on the backboard.

A FEW POINTS

Whether your children are attending their school, regional, state, or national fair, the one thing that will always be a constant is the backboard size. The backboard has to fit within the confines of the space allotted to the project itself, and that space is 30 inches deep, 48 inches wide, and 108 inches high including the table it is resting on. Since the backboard goes along the back of the project, its dimensions will be close to this size as well. Your child can do whatever she wants with this space, within reason of course, remembering that her apparatus (if she wants to show it), her report (has to be there), and her backboard all have to fit in this space. This is why we suggested using photographs for observations and apparatus. Space that otherwise would have been taken up by the apparatus can now be used to display photos of the apparatus at work, in various stages of assembly, and so on.

Because the backboard is physically at the back of the table, at least 30 inches from the reader, it has to show the work at a glance, so normal-size writing will not work. Large letters and printing will be needed, and even the paragraphs of information that will inevitably creep onto the board have to be lettered larger than usual. For computer generated text this should pose no problem, but for handwritten material it may cause some concern. Neat, large-style printing may be difficult for your child to do well enough for display. Alternatives are stencils, stick-on letters, or using a photocopier to enlarge the original print. Keep in mind that the end product needs to be neat, legible, and visible.

The final point to make before moving on to our actual backboard design is about the title. The title of a project is the main place where students can exercise a flair for the dramatic, or a pun, or a poem, or a lyric, or about anything cool in writing that they can think of. The title can be fun but still has to state exactly what the project is about. Two of the more popular science fair ways to do this are to make some sort of catchy title that says it all or make a witty statement followed by the title of the actual experiment. An example is the title of an experiment dealing with keeping fruit flies away from produce: "Shoo fly, don't bother me." It tells everything that is needed—the project deals with flies and keeping them away from you. Another example is the title of an experiment on the breaking down and cleaning up of discarded chewing gum: "Stuck on you: The bioremediation of chewing gum." Here the title has a cute start and then follows up by describing exactly what's going on. Another way to do this is to use subtitles. The hallmark of a good project is the creativity of its title. Creative titles catch the eye and the imagination, and show that the student was being creative. Judges look for more than catchy titles, but these are very popular with the public.

PURPOSE

The purpose written on the backboard will look much like the question the student has written in the report, but may need to be shortened, and in some ways expanded upon. The purpose is why the student did the research, and therefore it can incorporate parts of the original project idea before it became a question, especially if there are dramatic parts. The purpose on the backboard also has to be related to the title of the project and should tie in well with it. It is usually placed either at the top center (because all eyes are drawn to that point) or at the top left.

HYPOTHESIS

There may be some confusion between the purpose and the hypothesis, and many students will actually combine the two. Although this may seem like it works well, it is actually bad science fair form. The person (or judge) who looks at the backboard shouldn't have to search for the hypothesis, it should be separate from the purpose and stated clearly. It is important that the student make an educated guess, go out on a limb if you will, and defend his choice to the judges, his peers, and the general public. The distinct separation of the hypothesis will ensure that this happens to the fullest. The hypothesis should be placed directly under the purpose, wherever that was.

PROCEDURE

There are several different ways to lay out the procedure on the backboard, and all of them are correct. The student can take her procedure as it is written in her report, paraphrase it, and then put it right on the backboard. This is legitimate because it puts all the needed information onto the backboard. The trouble is, in this long form it may be difficult to pick out what was done when and whether the student followed proper experimental procedure. A simple and very popular way to rectify this is to put the procedure in point form. The drawback to the point form is that it assumes that one step logically follows another, and in some cases the steps may happen at the same time. Still, the point form can be used if you further identify the steps through time indexing. The student doesn't have to say exactly when each step was undertaken, but rather identify which step took place first, which second, whether two were done at roughly the same time, that sort of thing. When using point form, the student should use complete sentences and not be repetitive or start every sentence with "Next I did this." Either a diagram or picture of the working apparatus should accompany the procedure, even if that apparatus is displayed on the table in front of the backboard. The procedure can be placed on the backboard beneath the hypothesis or, if the hypothesis is on the left, at the top of the center.

MATERIALS

The materials used in the experiment may be the easiest things to put on the backboard because all that is needed is a list. The importance of this cannot be understated, though. If all the apparatus for the experiment is not present, this list is the only indication of what materials were used for the experiment. The only real problem with the materials list is where to physically locate it on the backboard. Many formal lab write-ups will have the materials list before the procedure, as they must be obtained before the experiment can be undertaken. In our case, space will decide where the materials list will go. If the procedure is on the center part of the backboard, then the materials will go to the left part under the hypothesis. This fulfills some of the formal lab write up requirements in that physically the materials are listed before the procedure.

RESULTS AND OBSERVATIONS

Raw data has no place on a backboard, not in table format, not in a chart, not anywhere. Only manipulated data should be displayed, with one or two sample calculations to show the judges that the student understood the process. Students can use the various graphs we discussed in the last chapter to show large chunks of data. These enhance the display and show the data in a way that can be interpreted quickly and easily. The logical location of this section is directly below the procedure, but many students find that more space than this is required to show all the necessary graphs and calculations. This section can continue up the right side of the backboard, as long as room is left for the final sections there.

SOURCES OF ERROR

We included sources of error in our conclusion and this can be done on the backboard as well. Some students put it ahead of the conclusion in its own place. This allows for a discussion of "possible" errors before the conclusive statements. If the student can give the impression that she knew the limitations of her experiment all along, rather than after she got some strange results, that may help in the judging section. Place this section directly below the observations or on the right side if the observations do not extend that far.

CONCLUSIONS

Unlike in their written report, students do not have all the space in the world for their conclusions, so they have to be careful what they put there. There isn't room for all the conclusive statements and their justifications on the backboard, so something has to give. That something is the major portion of the justifications. On the backboard, students can get away with statements such as, ". . . and this is (or isn't) supported by our results," or, "We found from our results that . . ." This gives a justification without having to state exactly where in the results it was coming from. The most important thing is for the students to put all of their conclusive statements on the backboard. Since the judges will interview the students, they can justify their conclusions orally. The conclusions will be the last thing on the right side of the backboard.

Now that we have a sense of what should be written on the backboard and where, we'll look at some other important considerations. If possible, have all backboard materials consistently prepared, either all handwritten or all computer printed. If the backboard mixes some computer-generated images, graphs, or text with handwritten materials, the judge might wonder, "Why didn't this student do everything on the computer? Maybe this student didn't do these himself." These are not thoughts that students want judges to have.

The International Science and Engineering Fair and the Canada Wide Science Fair have strict rules governing what a backboard can be made of. Certain things like corrugated cardboard or other such flammable materials are not allowed for the simple reason that it isn't safe to have four or five hundred of them in a room with students. The risk of a bad fire is too great, so the committees err on the side of caution. Acceptable materials are plywood, Sintra™ (a solid plastic), or any material with a reasonable fire rating. Check with your local or regional fair organizer. Taking the long view, a hinged plywood backboard will last for most of your child's science fair career, so it may be a good investment.

Of course, it doesn't do any good to make a backboard of fire-resistant material and then plaster papers all over it. Believe it or not, regulations cover this as well. Backboards may be painted but not covered with colored paper. Papers on the board have to be glued down completely, leaving no air space behind them.

The third thing to consider when looking at what to include on the backboard is whether or not your child's apparatus can be displayed. Remember the regulations of what can and cannot be shown at a science fair? Well this is where they apply big time. Safety considerations are paramount for a major fair. Refer to chapter 2 for the list of materials that cannot be displayed.

Since many things cannot be displayed in a science fair, photos will have to take their place. In fact, using well-planned photos on the backboard would be better than bringing items that can be displayed but not operated, as the apparatus would have no purpose other than to display what a photo could show just as well.

CASE STUDIES

It has been several chapters since our case study students have been mentioned, so this is a good place to bring you up-to-date.

ERIN

Erin has finished her project, and she has good results to report on and a reasonable set of conclusions. All she needs now is her written report and her backboard. She has kept good notes, so her written report wasn't too difficult. For completeness sake, Erin described her entire experimental process right from the first day she saw the alarm system at her father's office. She talked about the challenge she felt with the idea, the research she did as background, and the help she received from various agencies. When it came time to describe how she designed and wrote her program, she included samples of her programming at different stages and some dead ends she experienced. All of her raw code is listed in an appendix of her report. While she was putting everything together she remembered other aspects of her programming work that she had forgotten, and her knowledge of her project was strengthened. Her backboard preparation went well, again because of her good notes, but she reduced the amount of code presented, leaving that for her report. Her backboard contains mostly results, pictures of the experiment taking place, her conclusions, and what she found to be the fundamental problems with her research. She incorporated some drawings of burglar costumes, masks, gloves, and that sort of thing as a decoration of her backboard. Her title was, "Safe and Sound: Using a computer to enhance a burglar alarm." The next question to be answered was what she should display. After much thought (and pleading with her father), Erin decided to display a computer as part of her project, and with some help from her father she was also able to get a motion sensor for her demonstration. She felt (probably correctly) that a demonstration would enhance her project for the judges.

This presents an issue that we'll discuss in the next chapter—what should be displayed. Many students bring computers to the science fair for their graphics abilities when a picture would have done as well. Others bring them to display slide show presentations for the judging. But really, only those projects that deal specifically with computer hardware or software should actually display a computer, and then only if the computer is necessary to the project, not just to fill in the space in front of the backboard.

JOSEPH

Joseph also has to decide what he will display at the science fair. He has thought (quite correctly) that his soil samples will look good as a display and they show exactly what he has been doing. Small vials of the oils he used and a sample of the enzyme will also be included. Joseph didn't quite follow the process as we laid it down—he worked out what he would display from his experience with a science fair before he wrote his report and did his backboard. (This is not uncommon. Although we've laid out this process in the order that we think will be the easiest for students, if the student feels more comfortable doing several of these final steps in a different order, that's fine.)

Joseph's report was easier to write than he expected, because he kept good notes and followed a process. He used background material supplied from his uncle, properly referenced, and a series of diagrams showing how the enzyme he used aids the natural bacteria in the soil to break down the oil. His conclusions, you may remember from chapter 10, were fairly involved, yet he chose to expand on them by explaining how the sandy soil had more space and aeration available to the enzyme to allow it to do its work. Expanding on discovered knowledge with supporting research sources is an effective way of making conclusions even stronger.

His backboard was fairly sparse because his question, hypothesis, materials, and procedure were very straightforward. He decided that although his conclusions were pretty long he would use them for his backboard just as they were. His display of soils, oils, and enzyme would support the backboard pretty well and he had the room. He further decorated the backboard with some photos of various stages of his experiment and common places where oil spills occur. He thought for a long time about a title and finally came up with, "Oils well that ends well." Both the photos and the title proved to be a nice touch that added to Joseph's strong project.

• • • • • •

WILL

Will had a straightforward project and should have a reasonably straightforward time of writing up his report and doing his backboard. He started his report with where his idea came from and how he adapted it to his own use. He talks about his problems with his procedure and emphasizes his errors of time and lack of measurement in some cases. He mentions his first failed attempt, but doesn't quite explain why he repeated his experiment and why he didn't accept his first results.

Will created his backboard using nice plantlike colors interspersed with musical notes. He made it very decorative to add to the dramatic value of the project. He put up all his required materials but had few photos of his experiment in action. Even though he doesn't have a whole lot of stuff on the backboard, he still shortened his conclusions—an arguable point but one that makes little difference to his project. He decided, probably incorrectly, to display his plants and his sound damping apparatus. This has merit from a demonstration point of view but the plants have no meaning to the experiment anymore. They have been outside the experimental environment for over a week and now are just plants. In fact, the one plant that was shorter than the others due to the experiment is now just as tall. Will went with a straight description of his project for a title: "Does sound affect plant growth?"

Will's experience demonstrates a common problem for repeated experiments and for the students reporting them. If a student has repeated an experiment for what she feels was an anomalous result, then she should be able to explain why the result wasn't right. If she can't explain why the result wasn't right beyond the fact that it didn't match her hypothesis, then she either shouldn't have repeated the experiment or she shouldn't mention it in the report. That may seem dishonest but it isn't really; the student is reporting on the running of the experiment that got the results she was looking for. Another way to get around explaining the running of the experiment a second time is to call the first try a pilot project or a test run. Either way, if it can't be explained, it shouldn't be reported.

Science Fair Time

*These crazy-quilt displays of varied talent and
effort should be a highlight of the academic year.*
(Chiapetta and Foots 1984, p. 36)

The big day is finally here, no more experiments, no more write-ups,
just pure science fair day fun. Well not necessarily, but the more
planning that takes place ahead of time, the fewer headaches the student will have on fair day. We've finally reached the end of our science
fair project flowchart.

What can your child expect on the day of the fair? Most fairs,
depending on their size, last only one day, but as the student pro-
gresses up the ladder through regional and on to national and international fairs, the amount of time spent at the fair increases as well.
For example, high school fairs generally last one day, a regional fair
may last two, and the International Science and Engineering Fair and

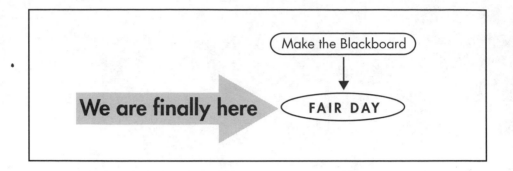

Canada Wide Science Fair may last as long as a week to ten days. The more participants and the higher the quality of the projects, the longer the fair will need to be. Of course, there are other considerations such as tours of local facilities, public viewing times, social events, and that sort of thing. Still it can be a daunting week.

THE SETUP

We had a plan for the other steps, and we need a plan for setting up the science fair project too. How complicated the plan will be depends on how complicated the display is going to be. There is a rule of thumb for these things: KISS—Keep It Short and Simple. Many science fairs are held during some part of the school day, during class time, lunch, or after school. Getting time off from work to help your child set up his project isn't easy, and some schools even frown on what they consider parental interference in this matter.

While you shouldn't interfere, you should try to be there at least for a few minutes, just to reassure your child. Sometimes a science fair can be a scary place. Those few minutes you spend checking on your child can mean a lot. As your child gets older and more experienced, he won't need you as much, so enjoy it while you can.

At any rate, your child needs to be able to set up her project by herself. Living by the KISS rule then becomes very important. It's a good idea to practice breaking down and setting up the project before the fair. Keep the backboard design simple and the amount and type of apparatus to a minimum, because the time for setting up is finite and ends all too soon.

Not only does the project have to be set up, but it also has to be moved to the fair site from its current resting place and actually arrive there in one piece. Boxes and packing materials might be needed. Ideally a student should arrive with only a backboard and some small pieces of apparatus. That is ideally—at our own school fair we have seen computers, large laser light shows, whole labs of glassware, even twenty-six experimental plants. The fact that the students can fit this stuff into their allotted project space is beside the point; most of it isn't required. Somebody had to carry this stuff into the school and return it later.

Pack the boxes at home the night before (no scurrying around on the morning of the fair), and pack in reverse order, so that the first part of the backboard or the apparatus comes out first. Label the boxes clearly, with lists if necessary (we hope not). Use good packing material to avoid breakage, and make sure that your child can unpack it. Keep the stuff to a minimum.

Once everything has arrived at the fair site, usually the student has to register his project and discover the location of his table. This is not as simple as it sounds; he may have to stand in line for a while holding his whole project. This is a prime time for you to be there for reassurance. Once registered, students can take all their materials to the table and start setting up. A fair official will come around to check the project for safety and size problems. If the rules have been followed for your local fair, this should be no problem. Students (and teachers) who did not bother to check the rules beforehand will pay the price here. After everything is set up, it's time to wait for the judges. Many regional fairs may have tours or activities for students to keep them occupied before the judging starts, otherwise, students can practice their presentations. At most local fairs the judging starts soon after setup, so they shouldn't have to wait too long.

HERE COME DA JUDGE

The first thing many judges will say is: "Tell me about your project." This is the student's chance to explain to the judging team everything she has done—in five minutes or less. She should be prepared to fit all the desired information into the time provided. She should be knowledgeable about her project and some of its background research. She should also be ready to defend her work, her conclusions, and her project as a whole, rather than depending on the judges' questions to bring out the important points. If a judge is familiar with the student's area of study, his questions will probably bring out the points anyway, but in some cases, especially at the local and regional level, the judges may not know much about the student's topic. The student has to be prepared for this eventuality and part of being prepared is to know and practice beforehand what should be included in an interview with a judging team.

Judges' questions may vary, but they all fall into the category of having the students defend their work. A common list of questions is found below, but remember, all judges are different, so it's difficult to provide even a general idea of what questions they might ask.

Why did you control this variable?
Why didn't you control that variable?

What do your findings mean?
Why do you think this was your result?
What can be done to correct this particular source of error?
What do you see as an application for your research?

Most of the time the judges will look at the project without the student being there, and they'll get a sense of what is happening by reading the backboard or summary. Then they'll think of informed questions to ask about each project. If the judges are prepared, then the students have to be doubly prepared.

Judges will frequently ask about the apparatus displayed, which is a good way to stir up questions and direct judges to specific areas. But if the judges feel that the apparatus wasn't necessary to display, the students will lose points. This goes back to what to display and what not to display. If you do not have a working model of your apparatus, then you do not need to display it, unless of course you have invented or developed something that you wish to show off. Common experimental materials are just that, common, so most scientific judges will have seen them before. The student will not impress these judges by having a complicated setup present on the table; it's more important to be able to defend the methods and conclusions.

It's also important to understand the judging criteria. Opinions vary somewhat for this, so we'll illustrate several different judging methods. Brian Hansen suggests that judges weigh their marks based on the student's creative ability, on the scientific thought that went into the project, on the student's thoroughness, and for neatness. His actual breakdown is 35 percent creativity, 35 percent scientific thought, 20 percent thoroughness, 10 percent neatness (Hansen 1983).

Another set of noted educators, Lawrence Bellipani, Donald Cotten, and Jan Marion Kirkwood, break their judging into what they see as several important criteria: creative ability, scientific thought, thoroughness, skill, and clarity (1984). Authors Eugene Chiappetta and Barbara Foots also provide five criteria: creativity, scientific thought, student understanding, skill, and student's own ability to explain the project (1984). Once again the themes remain the same and we can break them down into several common ones. Creativity, thoroughness and skill in undertaking the project, scientific thought used in the project, how well the students understand what they have done, and how well they have expressed and shown what they have done. From an official point of view, these criteria are a good sample of what you will see on the actual judging sheet. One from our regional fair is given in Appendix E. All these criteria speak to the students' work, where and how they got their ideas, how well they used scientific method, how complete their reports and backboards are, how well it is all put together, and finally how the students did in the interview itself. By doing the project themselves, the students have prepared themselves the best way they can for the judging process. Remember that above all else, if they have done it, written it down, studied it, and used it, then they should be able to defend it. There's no substitute for the intimate knowledge students gain from doing the projects themselves with only minimal help and lots of support.

The judges can spend as much as ten to fifteen minutes with each project depending on the number of projects to be judged. Usually there will be some time between projects for the judges to fill out their forms and talk over the project with their fellow judges. Judging groups tend to be made up of two or three judges, but at some local fairs the lack of qualified judges may mean one person is judging alone. After they have judged all their projects, the judges get together to discuss what they have done and to compare notes. Generally, the marks the students receive on the judging form are representative of how well they have done, but occasionally some judges mark harder than others, so discussing the projects among the judging groups ensures that the judging is as fair as possible. Such discussions also help eliminate the problems caused by single judges. Unfortunately, not all fairs practice this type of judges' discussion.

PRIZES, WHAT PRIZES?

Many science fairs award prizes for the best project in certain categories and divisions as well as the best in the whole fair. Prizes, of course, are not the only reason students should participate at the fair. Competing against their peers, the pride of completing a science project, and the academic value of the work are all important motivations. But winning a prize is fun too! Prizes are pretty rare at the local level—what the top projects win is usually the chance to compete at the next level of science fair competition, be it district, city-wide, regional, or state. Local fairs usually give out medals or ribbons at least. At the levels above the local fair, many prizes may be available, such as scholarships, software packages, plaques, or trips—the list depends on the generosity of the science fair donors and the hard work of the fair organizers.

The reality of the situation is that only a few projects will win awards, although many may receive ribbons, medals, or certificates. Being prepared to console your child is another important, but less pleasant, role that you may have to play. Remind your child that she has much to be proud of. Thinking up the idea, designing the experiment, and presenting it to the judges are all worthy accomplishments. Remembering that there is always a next year is an important consolation as well. Now that your child has experienced the science fair and seen the other projects, she should have a better idea of what it takes to win an award or progress to a higher level. If she doesn't grasp that right away, suggest a visit to the award winners. Some students even take notes on the differences between winning projects and their own.

THE PUBLIC VIEWING

Just because the judging is over does not mean that the fair is also over. Most fairs offer a chance for the public to see what the students have done. One of the nice things about these viewings is that the public will usually walk through the entire fair site and try to see everything rather than concentrating on just the winners. This means that the students may end up defending their projects by answering questions again and again. This makes each student feel that his project is important, which is one of the purposes of the public viewing. Another purpose is to let the general populace see what the school and the students are capable of doing. A third purpose is to give younger students an opportunity to see the older students' work and find out what science fairs and projects are all about. All of these purposes are wrapped up in a single public viewing session, which is pretty neat.

ACTIVITIES

Some science fairs incorporate other science- and engineering-related activities with the fair. Encourage your child to participate in these activities. The science and engineering activities vary from fair to fair, but for example, our own regional fair holds a Science Olympics. A Science Olympics is a multi-event competition based on teamwork and engineering and scientific principles. Students are given a problem to solve in teams, with a time limit and limited materials to do it. One of the most common examples is the "Save the egg drop." The students are provided with some tape, straws, and maybe a few paper clips, and they have to devise a way to have an egg survive a ten-foot drop. It's fun, but it also involves critical thinking skills, creates good peer relationships, and encourages students to think on their feet. By participating in these sorts of activities, your child will get more out of the whole science fair experience.

Another interesting aspect of the science fair, even at the local level, is the interest it generates in forming some sort of science club. The fact is that the science fair may have actually been spawned from just such a club in the 1920s (Asimov and Fredericks 1990), and now the reverse is true—some of these clubs actually spring out of the science fair. A good science club gives students opportunities to explore the world around them and to meet scientists and actually see them at work. Once again, participating is important for any child with an interest in science.

FINISHING TOUCHES

We'd like to leave you with a reminder that the science fair is your child's experience. You cannot live your children's lives for them, and in the same manner you cannot have this experience for them. They need your support, they may need you to be there, and they may need your help, but they do not need your interference. In science fairs, it's not whether you win or lose, it's how much you learned in the process.

But if you do want to get involved, turn to the next chapter where we'll tell you how parents can become involved in the actual setting up of a science fair and provide some alternatives to the science fair idea.

A Science Fair Primer

Though a successful Science Fair requires an enormous
amount of time and energy, the payoffs are impressive . . .
(Hansen 1983, p. 14)

Maybe your school hasn't had a science fair in a while, and maybe your school has never had a science fair. In either case, if you want one, you may have to organize it. We've discussed what a science fair can do for a student, but for there to be such an opportunity the fair has to be set up, and set up correctly. Anyone interested in being a science fair volunteer needs to have certain information, which we'll attempt to provide. We'll also present some alternatives to the traditional science fair.

LAYING OUT THE TASKS

As with any job or project, a science fair involves certain tasks, everything from recruiting judges to arranging space, from finding tables for the projects to getting a committee together. The first and most important task is getting the school on board. The school or schools in your area need to agree to support the science fair, because without the teachers' support, it quickly becomes impossible. In most cases, an initiative like this is started by the school or an individual teacher. If this is not the case, you should start by meeting with a science teacher to find out why. This meeting also serves as an introductory session between you and the local science teacher. Perhaps this meeting may lead to the formation of a science fair committee of some sort—who knows? Trying to organize such a fair without classroom backing will doom it from the beginning.

Just like the projects the students will be doing, you need to start early. The previous May or June would be best, but a September start date is not impossible. Many details have to be considered, and all the people you will be dealing with are volunteers. The earlier you start, the better chance you have of getting the fair organized in the time frame you want.

Just asking the right questions can identify many tasks. We will break down what has to be done by asking the simple questions usually used for writing newspaper articles: who, what, why, where, when, and how.

WHO

The first question is who will be involved with the fair. There are three separate and distinct groups: the committee, the students, and the judges. These groups need to be separate to avoid questions of favoritism or conflict of interest. The one exception to this is the chief judge, who is part of the committee and also a judge. But the chief judge does not actually judge, she organizes and advises.

The first step to this process is putting a committee together, for from this all other things will come. First, you need to decide what types of people should be on the committee. In the beginning, the committee can be fairly small, and you can add positions as the need arises. For starters, some type of scientist is necessary, mainly because this person has the contacts you need to find judges and will make a fine chief judge. Having a scientist on your committee will also legitimatize your efforts in the eyes of the general public. Second, you need educators for the obvious reason that you need the school's support and the school needs to know what is going on. Teachers bring the expertise of their profession and their experience in dealing with students, parents, and their own peers. You will find this help invaluable. The third group that needs to be represented is the parents. The parents are the ones who will make the process work at home, and they are experts when it comes to scheduling their own children and knowing what can and can't be expected of them selves and their children. A fourth group that is not essential but would be nice to have represented is the local business community. These people can help with donations, finding a site, and keeping the books, and they are a veritable fountain of ideas concerning educationally related events. Representatives from the local government or school board would also be nice. These people can provide the clout necessary to obtain certain sites, and they give legitimacy to the committee with respect to the community, have experience with the media, and may help with funding and donations as well. All of these people would be an asset to any committee, but the important thing to remember is that just as with the students and their ideas, interest is the key. A group of interested volunteers, no matter what their background, can accomplish a lot.

Once the committee is put together, certain responsibilities have to be handed out. Someone needs to recruit and organize the judges. This person is the chief judge and has this job and this job alone. A scientist or teacher is good for this position. They know whom to call and may have made connections with these people in the past. In fact, depending on the size of your fair, this may need to be more than one person. Next, someone has to handle the details of the fair itself, finding a site, getting tables, and that sort of thing. This person is the site manager and this is a good job for the businessperson or teacher. Once again, it depends on where the fair is being held, something we will discuss later. Then there needs to be someone in charge, a committee chairperson, mainly because someone has to call meetings and ensure that everything is being done when it should be done. Any of the committee members can fill this position, and in fact, it should rotate among the committee over time so that no one gets tired of the job. Someone needs to watch over the finances, a treasurer, who may also be involved with fundraising or gathering donations. The business representative or local government representative would be good for this position. They have the contacts needed for this type of work, and may already be doing it as some function of their careers. Other positions will manifest themselves as the fair grows; publicity chairperson, secretary, registrar, and so on. Starting out though, there only needs to be a bare-bones organization, and the positions outlined above will fulfill most of the needs of any new science fair committee.

A second "who" that needs to be considered is who will be the judges. Employees from local institutions such as hospitals, colleges, universities, or even public utilities can form the core of a judging group. Other people in your community that may make good judges include professionals, scientists, retired teachers, and the like. Try to avoid using teachers at the local school as judges because they know the students too well and may not be able to judge their work impartially. In larger areas, teachers sometimes trade off, with one school judging another school's fair projects. In the end, you will have to use whomever you can get, and the most important thing is to make sure these people understand what they are doing and that there are enough of them. Ideally, there should be one judging group for each six to eight projects, and each judging group should contain two or more people. This may not be possible, but it is the ideal.

The final "who" is who will be participating in the fair? What students will be participating, what grades, what schools, how far an area will it all encompass? If you are dealing with more than one school, the whole process becomes more difficult. It may be a good idea to start with only one school. The next issue is to decide which grades will participate. A classic science fair usually involves the junior and senior high grades, grades seven to twelve. Grade seven is the starting age for competition at the higher levels of a science fair. You can include the younger grades though, and there are several reasons why you should. The experience that younger students gain by attending a science fair cannot be understated. They may not have the best projects in their first year, but they will gain a better idea of what the fair is about and what good projects look like. The second reason is that at about the fourth-grade level, students should start experiencing practical science. They are ready to handle it intellectually and fewer expectations are put on them. The main argument against the inclusion of the elementary grades is the issue of competition. Many teachers and educational researchers feel that competition in the younger grades isn't healthy; they will face it soon enough as it is. Even students at this level recognize the need to "avoid a competitive overtone" (Pike 1993). However, there are ways to incorporate the work of elementary students in the science fair without having them compete. Having a separate, noncompetitive science exposition for the elementary students would be one way to accomplish this. The great thing is that the planning necessary for such a day overlaps the planning for the rest of the fair, meaning major dividends for not much extra work.

WHAT

The "what" question is what kind of fair do you want to have, what needs must it meet, and what are you trying to accomplish by doing the fair. This refers back to who is in the fair and ahead to why the fair exists. Many different types of fairs can be undertaken, and some events can be undertaken in concert with a science fair, which will make the whole process that much better. For starters, you can have a science fair that does not emphasize competition, meaning that all projects are praised for their merits, and no one project is singled out for special recognition. Or you can have a competitive fair in concert with a noncompetitive exposition. Either way will allow lower grades to participate without the worries of competition. Bob Burtch, a fifth-grade teacher in Illinois, sees a noncompetitive fair as essential to his students' learning. "The science fair at our school is designed as a teaching tool rather than a contest, and my aim is to involve and enrich all students in my fifth-grade class." (Burtch 1983)

The separation of competition from the science ensures that the science does not get lost and that the student's learning takes precedence, which is important for all grades. In junior and senior high grades though, the reality is that competition exists, and preparing students for real life is one of the functions of these grade levels. The possibility of winning an award, a prize, or the right to represent their school at a higher level of science fair competition is one of the motivations for students to participate. Noncompetitive fairs don't offer this motivation.

Here are some alternatives to a science fair that you can explore if the idea of a typical competitive science fair doesn't sound good for your situation. One is the consumer fair (remember our example of a consumer-based project). A consumer fair involves students testing consumer products to see which one is the best for the lowest cost. Basically, they are re-creating the popular commercial tests that we see quite frequently. The interesting thing about consumer fairs is that the ideas for the projects are a little easier to find and the form the project takes is already decided for the students so more time is available for actually designing and undertaking the experimental part. Donald Nelson, an educator from Illinois, sees many benefits to the consumer fair as opposed to the normal science fair. "The activities provide reinforcement of skills taught in the science classroom and help students become knowledgeable and independent consumers. Most important, consumer fairs illustrate that science and the methods of science are relevant to students' everyday lives." (Nelson 1986, p. 56)

The interesting.thing about this alternative is that it uses the same format as the science fair, and therefore setting it up requires about the same procedure as a science fair. This provides the possibility of having alternate years for science fairs and consumer fairs just to give the students something new to do.

A second alternative to the science fair is the science showcase night. James Scarnati, William Kent, Lucia Falsetti, and John Golden were tired of the science fair scene and had an objective of having, "all students be winners" (Scarnati et al. 1992). Their showcase night involves students creating displays, setting up presentations, and designing activities on many general scientific principles. The interesting thing about it is that the showcase focused more on the guests who would be attending the event. All of the work the students put into their own projects wasn't about the projects themselves but about demonstrating scientific principles to the general public and showing how things work. Students learn something better when they teach it, so this is a terrific benefit. And although the participants don't compete against each other, they can see who thought up the best display. Science showcase night seems like an interesting spin on the old science fair idea. Detailed timelines that highlight the differences between a showcase night and a science fair are found in the article by James Scarnati, William Kent, Lucia Falsetti, and John Golden (Scarnati et al. 1992).

Some schools hold science days and open houses to offer a change from the competitive science fair. A science day involves the students in the school in a day of science activity, based at stations set up around the school. Each station may include guessing games, design activities, Science Olympics events, science word games, or small field trips (Simms 1993). Wayne Simms, an educator from Newfoundland, uses teachers, senior students, parents, and visiting scientists to run the whole show. The students are the main participants and the effect is like a large science carnival. An open house is much the same, but it involves the public along with the students. An open house can just be a walk through the facilities available at the local school, but activities and stations like those mentioned above will add to the educational opportunities. Both of these events involve planning much like a science fair, with just a few twists and turns along the way.

One final alternative to the science fair is a science exposition. Unlike the noncompetitive showcase night or science day, the science exposition can easily contain a traditional science fair competition, but it will also contain other activities that will appeal to students who are not interested in competing. This is the best of both worlds—incorporating the high educational ideals of the noncompetitive events with the ability to compete at regional, state, and national fairs. Debbie Silver, a Louisiana middle school teacher, created a science exposition at her school after learning that some of her students wanted to compete and some did not. Her "Science Expo" includes a myriad of activities for the public, students, and teachers alike—everything from a traditional science fair in one part of the building, "to share fairs, class demonstrations, the Invention Convention, the Family Physics Fun Festival, and the Family Science Olympiad" in other parts of the building. This is not for the faint of heart though, and even Ms. Silver herself suggests "neophytes start small" (Silver 1994). But in the end, this gives you the best opportunity to please all of the people for at least some of the time. Once again much of the planning for this type of event is similar to that for a science fair, with committee members coordinating each individual part of the expo.

WHY

Identifying the main reason we are holding a science fair will help determine some of the decisions we may have to make along the way. If it's for education, then center it on education—take away the competitive aspect and try to get as much teaching and learning into the fair as possible. But if it's for the students, then center it on them. Address the questions of motivation and doing what is best for the students. Decide what process can be used to develop an interest in science within the students themselves. Plan supplemental and preparatory activities for the students so that science fair day involves more than just the presentation of their projects. And we have to look beyond the fair to how well we are preparing them for real life. Essentially this decides the competition debate for us.

WHERE

Where to hold the fair is the next question to consider. "Where" is a relatively important question that is dependent on many of the other questions in this chapter. The number of students participating, when in the year it will be held, and what type of fair it is going to be are all questions that have to be answered before we can consider where the fair will be held. The number of different schools, expected projects, and projects that may require electricity all have to be considered. A school gym is a great place to hold a fair. It is convenient for the students and teachers, and there are extra rooms available for judging and organization. If more than one school is involved, the fair can be rotated from gym to gym each year so that the work and the honor of hosting the fair is shared equally. If the fair will involve a lot of projects, the gym may be too small, and if many projects require electricity, the gym may not have enough outlets. You also have to consider the number of tables and chairs needed. Other possible venues include public halls or local roller skating or ice skating rinks, but depending on the time of year, these may not be available either. You'll have to plan your fair around the availability of a suitable space, and you may have to limit your fair in some way to accommodate that space.

WHEN

You may think that the best time to hold the science fair is at the end of the school year, but that is not always the case. If your fair is planning to send students on to the regional or state fair, then the timing of your fair is dependent on the district or regional fair's date. Your winners have to be able to register before the deadlines have passed. You also have to consider things such as spring break, peak exam times, and the availability of your preferred fair site. This is where your teacher representative will come in handy. He can help you with his knowledge of what activities are coming up for students and how the school feels about potential fair dates. All these factors will determine the timing of your fair. Just don't make it too early in the year because the students need time to do their projects.

HOW

The biggest questions of all is "how." This includes many site-oriented items such as tables, electricity, how many people can safely fit into the site, medals or ribbons, awards ceremony, public viewing, school tours, other planned activities for the students, chaperones for the students, and basically anything else that is needed to run a fair. If you're holding the fair in the school gym, getting tables from the school should be easy, and electricity shouldn't be a problem either as long as your school (and gym) is fairly new. Just make sure to use surge protectors on your extension cords. The number of students that will be attending is always an issue, depending on how many schools will be involved and how many grades are invited to participate. Another problem here is the question of participation. Some schools make the science fair project mandatory for the students, which ensures that there are lots of projects and that all students are participating. Other schools leave it up to the students themselves but give incentives such as academic grades and assignment relief for those who do the projects. Illinois' State Board of Education teacher's handbook on science fairs goes to some length to present both sides of this argument. Their arguments are encapsulated below.

VOLUNTARY PARTICIPATION

Pros	Cons
Easiest	Overall success of the fair may suffer
Small numbers	Need to promote
Positive student-teacher interactions	Lack of student involvement
Few parental complaints	Lack of other teacher involvement
Positive attitudes during fair	Lack of administration involvement
No management problems	Possible lack of parental involvement

MANDATORY PARTICIPATION

Pros	Cons
Whole school involvement	Management problems
All cons can be overcome with good planning	Location
	Judges
	Parental complaints
	Too much parental involvement
	Poor student-teacher interaction time

Source: Riggins 1985

This decision has to be based on the viewpoint of the school and how the pros and cons of each choice affect the science fair as a whole.

The question of prizes was discussed earlier, but some token of participation should be given out. Some fairs give ribbons to all the participants, and then use a plaque, trophy, or medal to single out the best projects. Others just single out the best projects; the choice is up to the organizers. If you are giving out awards, then you need to hand them out in some sort of ceremony. This can take place during the public viewing, as a separate event, in the school after the fair, or whatever. If a ceremony is to be held, it should be as close to the judging as possible, but beyond that, the timing of everything else is up to the committee.

Public viewing and school tours serve to inspire future science students, motivate students who chose not to do projects, show donors and sponsors of the fair what it is all about, and to help inspire volunteers for the next science fair committee. Public viewing has to happen after the judging, so that you don't risk having a project damaged by the public before it gets judged.

The last thing to plan for is what to do with students if they have any free time. Activities for the students become important if there is going to be time when they are not being judged and have nothing else to do. Science booths, science games, guest speakers, Science Olympics, or even science fair scavenger hunts are all fine examples of student activities.

WORDS TO THE WISE

Keep it short and simple for now. The KISS rule returns, especially if this is your first foray into organizing a science fair. On your first try, stick to the basics: a suitable location, tables for the projects, power for those who need it, judges to judge, and simple ribbons awarded during the public viewing. The first fair that we set up at a local school accomplished all these tasks and netted about fifty projects and a couple of hundred visitors. But the next year, when we had more experience, our fair had a hundred projects and had close to a thousand visitors. Keep in mind that the bigger the fair, the bigger the organization. Cities that host the Canada Wide Science Fair bid for the honor five years in advance and very few of those days are wasted, but they have to deal with more than six hundred participants from all over Canada who will stay in the city for a whole week. This is quite different from the local fair planning described here.

Remembering the answers to the "why" question we asked earlier should help you keep focused when you are planning events of this type. Remember why you are doing this, and for whom, so you don't lose sight of your goal. A science fair isn't about you, or the businesses that donated prizes, or the committee that organized the fair itself, or the teachers and school that supported the idea. It's about the students and the work they have done. All these other people are important and deserve thanks and recognition, but in the end all the work and effort is for the students.

Conclusions

*This completes the yearly cycle which
began the previous spring.*
(Bombaugh 1987, p. 18)

The journey is over. Hopefully you have been successful in helping your child with her science fair project and everything is winding down. And hopefully your child has done well at the science fair, or at least learned something. You also should have learned something, which is that the science fair project was not for the faint of heart, but that sticking it out holds many rewards beyond the ribbons and medals. As science fair organizers, we thank you for the work you've done with your child and applaud your efforts to not interfere with her project.

Let's finish by wrapping up the stories of the three science fair students, and tell you how they did.

CASE STUDIES

ERIN

Erin had some success at her local fair, qualifying to attend their regional fair. Her judging went well and she impressed those she met at the local fair. At the regional fair the judging was tougher, and although her innovation was quite good, it was not good enough to take first place, being beaten out by a project that dealt with artificial intelligence. She still felt that it was all worthwhile and by competing at the higher level she found some ideas for her next fair, which she is now looking forward to.

• • • • • •

JOSEPH

Joseph also competed at his local fair, but was unable to move up to the regional fair. Although his project was well done and experimentally sound, his backboard and presentation to the judges wasn't top-notch. Complacency caused this to happen, possibly because of his previous science fair experience, possibly because his teacher graded his project before the fair and gave him a high grade. In either case, he learned from his mistakes this year. Hopefully they won't be repeated next year.

• • • • • •

WILL

Will's class was given the option of competing in the school fair or not, and given the age of the students and other factors we talked about in the last chapter, that was a pretty good idea. Will did not want to compete with his project after he saw the level of some of the other projects in his class. His project was well received by his teacher though, mainly because it was one of the projects she had suggested. He did well enough, and he also has an idea of how to do better next year. This is a positive thing in many ways, especially since in two years he'll be going to a high school that makes science fair participation mandatory.

We hope that the science fair project process is not quite as intimidating as it once was, and we also hope that you and your child have chosen to continue to participate. A science fair is only as good as the students who participate in it, and science fairs will only survive if parents like you take an interest in them. Good luck, and good science.

APPENDICES

APPENDIX A

These are the Intel ISEF affiliated fairs for 1998–99. An updated list can be found on the Science Service Website at www.sciserv.org.

United States

Alabama

Birmingham: Central Alabama Regional Science and Engineering Fair—Physical Sciences and Life Sciences

Decatur: North Alabama Regional Science and Engineering Fair—Life Sciences and Physical Sciences

Huntsville: Alabama State Science and Engineering Fair

Livingston: West Alabama Regional Science Fair

Mobile: Mobile Regional Science Fair

Talladega: Northeast Alabama Regional Junior Academy Science Fair
Troy: Southeast Alabama Regional Science Fair

Alaska
Anchorage: Alaska Science and Engineering Fair

Arizona
Prescott: Northern Arizona Regional Science and Engineering Fair
Sierra Vista: SSVEC's Youth Engineering and Science Fair
Tempe: Central Arizona Regional Science and Engineering Fair
Tucson: Southern Arizona Regional Science and Engineering Fair

Arkansas
Arkadelphia: South Central Arkansas Regional Science Fair
Batesville: North Central Arkansas Regional Science Fair
Conway: Arkansas State Science Fair
Fayetteville: NW Arkansas Regional Science and Engineering Fair
Hot Springs: West Central Regional Science Fair
Jonesboro: Northeast Arkansas Regional Science Fair
Little Rock: Central Arkansas Regional Science Fair
Magnolia: Southwest Arkansas Regional Science Fair
Monticello: Southeast Arkansas Regional Science Fair

California
Alhambra: Alhambra Science and Engineering Fair
Fresno: Central California Regional Science, Mathematics, and
    Engineering Fair
Palos Verdes Peninsula: Allied Signal Regional Science and Engi-
    neering Fair
Pleasanton: Tri-Valley Science and Engineering Fair

Sacramento: 1999 Sacramento Region K–12 Science and
    Engineering Expo
San Diego: Greater San Diego Science and Engineering Fair
San Francisco: San Francisco Bay Area Science Fair, Inc.
San Jose: Santa Clara Valley Science and Engineering Fair

Colorado
Alamosa: San Luis Valley Regional Science Fair, Inc.
Brush High School: Morgan-Washington Bi-County Science Fair
Colorado Springs: Pikes Peak Regional Science Fair
Fort Collins: Colorado Science and Engineering Fair
Grand Junction: Western Colorado Science Fair
Greeley: Longs Peak Science and Engineering Fair
Sterling: Northeast Colorado Regional
Trinidad: Spanish Peaks Regional Science Fair

Connecticut
Hamden: Connecticut State Science Fair
Ridgefield: Science Horizons, Inc., Science Fair and Symposium

Florida
Arcadia: Heartland Regional Science and Engineering Fair
Bradenton: GTE-Manatee Regional Science and Engineering Fair
Bushnell: Sumter County Regional Science Fair
Crystal River: Citrus Regional Science and Engineering Fair
Fort Myers: Thomas Alva Edison East Regional Science Fair
Fort Walton Beach: Southeast and Northeast Panhandle Regional
    Science and Engineering Fair
Fort Pierce: Treasure Coast Regional Science and Engineering Fair
Gainesville: State Science and Engineering Fair of Florida

Jacksonville: Northeast Florida Kiwanis Regional Science and
    Engineering Fair
Kissimmee: The Osceola Regional Science Fair
Lake City: Suwannee Valley Regional Science and Engineering Fair
Lakeland: Polk County Regional Science and Engineering Fair
Leesburg: Lake County Regional Science and Engineering Fair
Marianna: Chipola Regional Science and Engineering Fair
Melbourne: South Brevard Science and Engineering Fair
Merritt Island: Brevard Intracoastal Regional Science and
    Engineering Fair
Miami: South Florida Science and Engineering Fair: I
New Port Richey: ASCO Regional Science and Engineering Fair
Ocala: Big Springs Regional Science Fair
Orlando: Orange County Regional Science and Engineering Fair
Palatka: Putnam Regional Science and Engineering Fair
Panama City: Florida Three Rivers Regional Science and
    Engineering Fair
Pensacola: West Panhandle Regional Science and Engineering Fair
Plantation: Broward County Science Fair
Quincy: West Bend Regional Science and Engineering Fair
Saint Augustine: River Region East Science Fair
Saint Petersburg: Pinellas Regional Science and Engineering Fair
Sanford: Seminole County Regional Science, Math, and Engineering Fair
Sarasota: Sarasota Regional Science, Engineering, and Technology Fair
South Daytona: Tomoka Region Science and Engineering Fair
Spring Hill: Hernando County Regional Science and Engineering Fair
Stuart: Martin County Regional Science and Engineering Fair
Tallahassee: Capital Regional Science and Engineering Fair
Tampa: Hillsborough Regional Science Fair
Titusville: Brevard Mainland Regional Science and Engineering Fair

Vero Beach: Indian River Regional Science and Engineering Fair

West Palm Beach: Palm Beach County Science and Engineering Fair

Georgia

Albany: Darton College/Merck Regional Science Fair

Athens: Georgia State Science and Engineering Fair

Atlanta: Atlanta Science and Mathematics Congress

Augusta: Central Savannah River Area Science and Engineering Fair

Brunswick: Coastal Georgia Regional Science and Engineering Fair

Ellijay: Gilmer County High School Fair

Griffin: Griffin RESA Regional Science Fair

Milledgeville: Georgia College and State University Regional Science
and Engineering Fair

Savannah: First Congressional District Science and Engineering Fair

Warner Robins: Houston Regional Science and Engineering Fair

Hawaii

Ewa Beach: Leeward District High School Science Fair

Honolulu: Hawaii Association of Independent Schools Science and
Engineering Fair

Honolulu: Hawaii State Science and Engineering Fair

Kailua: Windward District Science and Engineering Fair

Kapaa: Northeast Kauai Regional Science and Engineering Fair

Wailuku: Maui Schools Science and Engineering Fair

Illinois

Chicago: Chicago Public Schools Student Science Fair

Edwardsville: Illinois Junior Academy of Science Region XII SF

Macomb: Heart of Illinois Science and Engineering Fair

Springfield: Illinois Junior Academy of Science Region X Science Fair

Indiana
Angola: Northeastern Indiana Tri-State Regional Science Fair
Bloomington: South Central Indiana Regional Science and
    Engineering Fair
Evansville: Tri-State Regional Science and Engineering Fair
Fort Wayne: Northeastern Indiana Regional Science and Engineering Fair
Greencastle: West Central Indiana Regional Science and Engineering Fair
Hammond: Calumet Regional Science Fair
Hanover: Southeastern Indiana Regional Science Fair
Indianapolis: Central Indiana Regional Science and Engineering Fair
Indianapolis: Hoosier Science and Engineering Fair
Muncie: East Central Indiana Regional Science Fair
South Bend: Northern Indiana Regional Science and Engineering Fair
West Lafayette: Lafayette Regional Science and Engineering Fair
Westville: Northwestern Indiana Science and Engineering Fair

Iowa
Ames: Iowa State Science and Technology Fairwa Energy Center
Bettendorf: QCSEF and Symposium: Greater QC Metro Area
Bettendorf: QCSEF and Symposium: Davenport Metro Area
Cedar Rapids: Eastern Iowa Science and Engineering Fair
Indianola: South Central Iowa Science and Engineering Fair

Kansas
Wamego: Wamego Regional Science and Engineering Fair

Kentucky
Bowling Green: Southern Kentucky Regional Science Fair
Lexington: Central Kentucky Regional Science Fair
Louisville: Louisville Regional Science Fair

Louisiana

Baton Rouge: Louisiana Science and Engineering Fair

Baton Rouge: Region VII—Science and Engineering Fair

Bossier City: Bossier Parish Community College Louisiana Region I
    Science and Engineering Fair

Houma: Terrebonne Parish Science Fair

Lafayette: University of Southwestern Louisiana Region VI Science Fair

Lake Charles: Louisiana Region V Science and Engineering Fair

Lutcher: St. James Parish Science Fair

Monroe: Louisiana Region III Science and Engineering Fair

Natchitoches: Louisiana Region IV Science Fair

New Orleans: Greater New Orleans Science and Engineering Fair

Ruston: Louisiana Region II Science and Engineering Fair

Thibodaux: Louisiana Region X Science and Engineering Fair

Maryland

Baltimore: Baltimore Science Fair; Morgan State University Science-
    Math-Engineering Fair

Frederick: Frederick County Science and Engineering Fair

Frostburg: Western Maryland Science Expo

Gaithersburg: Montgomery Area Science Fair

Largo: Prince George's Area Science Fair

Massachusetts

Amherst: Massachusetts Region I Science Fair

Bridgewater: Massachusetts Region V Science Fair

Cambridge: Massachusetts State Science Fair

Fall River: Massachusetts Region III Science Fair

Somerville: Massachusetts Region IV Science Fair

Worcester: Massachusetts Region II State Science Fair

Michigan

Ann Arbor: Southeastern Michigan Science Fair
Benton Harbor: Southwest Michigan Science and Engineering Fair
Big Rapids: West Central Michigan Science and Engineering Fair
Detroit: Science and Engineering Fair of Metropolitan Detroit, Inc.
Flint: Flint Area Science Fair
Port Huron: St. Clair County Science and Engineering Fair
Saginaw: Saginaw County Science and Engineering Fair

Minnesota

Bemidji: Northern Minnesota Regional Science Fair
Duluth: Minnesota Academy of Science State Fair
Duluth: Northeast Minnesota Regional Science Fair
Mankato: South Central Minnesota Regional Science and
    Engineering Fair
Mankato: Southwest Minnesota Regional Science and Engineering Fair
Minneapolis: Twin Cities Regional Science Fair
Minneapolis: St. Paul Science Fair
Minneapolis: Western Suburbs Science Fair
Moorhead: Western Minnesota Regional Science Fair
Rochester: Rochester Regional Science Fair
Saint Cloud: Central Minnesota Regional Science Fair and Research
    Paper Program
Winona: Southeast Minnesota Regional Science Fair

Mississippi

Biloxi: Mississippi Region VI Science and Engineering Fair
Booneville: Mississippi Region IV Science Fair
Greenville: Mississippi Region III Science and Engineering Fair
Hattiesburg: University of Southern Mississippi Region I Science
    and Engineering Fair

Jackson: Mississippi Region II Science and Engineering Fair;
    Mississippi State Science and Engineering Fair
Mississippi State: Mississippi Region V—N & S Science and
    Engineering Fair
University: Mississippi Region VII Science and Engineering Fair

Missouri
Cape Girardeau: Southeast Missouri Regional Science Fair
Hillsboro: Mastodon Park Science/Art Connection Fair
Jefferson City: Lincoln University Regional Science Fair
Joplin: Missouri Southern Regional Science Fair
Kansas City: Greater Kansas City Science and Engineering Fair
Manchester: Greater St. Louis Science Fair
Rolla: South Central Missouri Regional Science and Engineering
    Fair
Saint Joseph: Mid-America Regional Science and Engineering Fair
Springfield: Ozarks Science and Engineering Fair
St. Peters: St. Charles-Lincoln County Regional Science and
    Engineering Fair

Montana
Billings: Deaconess Billings Clinic Science Expo
Butte: Southwest Montana Regional Science and Engineering Fair
Great Falls: Montana Region II Science and Engineering Fair
Havre: Hi-Line Regional Science and Engineering Fair: MSU-Northern
Missoula: Montana Science Fair

Nebraska
Hildreth: Central Nebraska Science and Engineering Fair
Nebraska City: Greater Nebraska Science and Engineering Fair

Nevada
Elko: Elko County Science Fair
Reno: Western Nevada Regional Science Fair

New Hampshire
Lincoln: New Hampshire State Science and Engineering Fair

New Jersey
Jersey City: Hudson County Science Fair
Randolph: North Jersey Regional Science Fair: Senior Division
Trenton: 44th Mercer Science and Engineering Fair

New Mexico
Albuquerque: Northwestern New Mexico Regional Science and
    Engineering Fair
Albuquerque: National American Indian Science and Engineering Fair
Farmington: San Juan New Mexico Regional Science and
    Engineering Fair
Grants: Four Corners Regional Science and Engineering Fair
Las Cruces: Southwestern New Mexico Regional Science and
    Engineering Fair
Las Vegas: Northeastern New Mexico Regional Science and
    Engineering Fair
Portales: Southeastern New Mexico Regional Science and
    Engineering Fair
Socorro: New Mexico Science and Engineering Fair

New York
Brooklyn: Polytechnical University NYC Board of Education Math,
    Science, and Technology Fair

Brooklyn: New York City Board of Education/City University of
New York Science Fair—Biological Sciences
Farmingdale: Long Island Science and Engineering Fair: VI
Poughkeepsie: Dutchess County Regional Science Fair
Syracuse: Greater Syracuse Scholastic Science Fair
Troy: Greater Capital Region Science and Engineering Fair
Utica: Utica College Regional Science Fair

North Carolina
Salisbury: Southwest North Carolina Regional Science, Mathematics,
and Engineering Fair

North Dakota
Devils Lake: North Central North Dakota Regional Science and
Engineering Fair
Dickinson: Southwest North Dakota Regional Science and
Engineering Fair
Fargo: Southeast North Dakota Regional Science and Engineering Fair
Grand Forks: Northeast North Dakota Regional Science and
Engineering Fair
Jamestown: Southeast Central North Dakota Science and
Engineering Fair
Mandan: Southwest Central North Dakota Regional Science and
Engineering Fair
Minot: Northwest Central North Dakota Regional Science Fair
Valley City: North Dakota State Science and Engineering Fair
Watford City: Northwest North Dakota Regional Science Fair

Ohio
Archbold: Northwest Ohio Science and Engineering Fair
Athens: Southeastern Ohio Regional Science and Engineering Fair

Canton: Ohio Region XIII Science and Engineering Fair
Cleveland: Northeastern Ohio Science and Engineering Fair
Columbus: Buckeye Science and Engineering Fair
Dayton: Dayton Science and Engineering Fair

Oklahoma
Ada: Oklahoma State Science and Engineering Fair
Alva: Northwestern Oklahoma State University Regional Science
    Fair
Bartlesville: Bartlesville District Science Fair
Edmond: Central Oklahoma Regional Science Fair
Lawton: Cameron University Regional Science Fair
Miami: Northeastern Oklahoma A&M Science and Engineering Fair
Muskogee: Muskogee Regional Science and Engineering Fair
Oklahoma City: Oklahoma City Regional Science and
    Engineering Fair
Seminole: East Central Oklahoma Regional Science and
    Engineering Fair
Tulsa: Tulsa Regional Science and Engineering Fair
Wilburton: Eastern Oklahoma Regional Science and Engineering Fair

Oregon
Forest Grove: Northwest Science Exposition
Gold Beach: Southwestern Oregon Regional Science Exposition

Pennsylvania
Carlisle: Capital Area Science and Engineering Fair
Lancaster: Lancaster Science and Engineering Fair
Philadelphia: Delaware Valley Science Fair
Pittsburgh: Pittsburgh Regional Science and Engineering Fair

Reading: Reading and Berks Science and Engineering Fair
York: York County Science and Engineering Fair

Rhode Island
Warwick: Rhode Island Science and Engineering Fair

South Carolina
Charleston: Low Country Science Fair
Clemson: Anderson-Oconee-Pickens Regional Science Fair
Columbia: South Carolina Region II Science and Engineering Fair
Florence: Sand Hills Regional Science Fair
Greenville: Upstate South Carolina Region I Science Fair
Spartanburg: Piedmont South Carolina Region III Science Fair

South Dakota
Aberdeen: Northern South Dakota Science and Math Fair
Albuquerque: National American Indian Science and Engineering Fair
Brookings: Eastern South Dakota Science and Engineering Fair
McLaughlin: Northwest Area Schools Regional Science and
    Engineering Fair
Mitchell: South Central South Dakota Science and Engineering Fair
Rapid City: High Plains Regional Science and Engineering Fair

Tennessee
Chattanooga: Chattanooga Regional Science and Engineering Fair
Cookeville: Cumberland Plateau Regional Science and Engineering Fair
Jackson: West Tennessee Regional Science Fair
Knoxville: Southern Appalachian Science and Engineering Fair
Memphis: Memphis-Shelby County Science and Engineering Fair
Nashville: Middle Tennessee Science and Engineering Fair

Texas

Amarillo: High Plains Regional Science Fair

Austin: Austin Area Fair and Festival

College Station: Brazos Valley Regional Science and Engineering
Fair

Dallas: Dallas Morning News—Toyota Regional Science and
Engineering Fair

El Paso: Sun Country Science Fair

Fort Worth: Fort Worth Regional Science Fair

Harlingen: Rio Grande Regional Science Fair

Houston: Science Engineering Fair of Houston

Kilgore: East Texas Regional Science Fair

Laredo: Laredo Independent School District Science Fair

Laredo: United Independent School District Regional Science Fair

Lubbock: South Plains Regional Science and Engineering Fair

Odessa: Permian Basin Regional Science Fair

San Angelo: District XI Texas Science Fair

San Antonio: Alamo Regional Science and Engineering Fair; Alamo
Regional Science and Engineering Fair III

Victoria: Texas Mid-Coast Regional Science and Engineering Fair

Waco: Central Texas Science and Engineering Fair

Wichita Falls: Red River Regional Science and Engineering Fair

Utah

Brigham City: Box Elder Science and Engineering Fair

Cedar City: Southern Utah Science and Engineering Fair

Clearfield: North Davis Area Science and Engineering Fair

Ogden: Utah Science and Engineering Fair

Provo: Central Utah Science and Engineering Fair

Virginia
Arlington: Northern Virginia Science and Engineering Fair
Central VA Community College: Central Virginia Regional
    Science Fair
Charlottesville: Piedmont Regional Science Fair
Dublin: Blue Ridge Highlands Regional Science Fair
Fairfax: Fairfax County Regional Science and Engineering Fair
Harrisonburg: Shenandoah Valley Regional Science Fair
Roanoke: Western Virginia Regional Science Fair
Manassas: Prince William-Manassas Regional Science Fair
Leesburg: Loudoun County Regional Science and Engineering Fair
Sterling: Virginia State Science and Engineering Fair
Virginia Beach: Tidewater Science Fair

Washington
Bremerton: Washington State Science and Engineering Fair
Kennewick: Mid-Columbia Regional Science and Engineering Fair
Tacoma: South Sound Regional Science Fair

Washington, D.C.
Washington: District of Columbia Mathematics, Science, and
    Technology Fair

West Virginia
Elkins: Eastern Regional Science Fair
Keyser: West Virginia Eastern Panhandle Regional High School
    Science Fair
Montgomery: Central West Virginia Regional Science and
    Engineering Fair
West Liberty: West Virginia State Science and Engineering Fair
West Liberty: West Liberty State College Regional Science and
    Engineering Fair

Wisconsin
Milwaukee: Southeastern Wisconsin Science and Engineering Fair
Superior: University of Wisconsin—Superior Science Fair

Wyoming
Cheyenne: Southeast Wyoming Regional Science Fair
Greybull: Northern Wyoming District Science Fair
Laramie: Wyoming State Science Fair

U.S. Territories

American Samoa
Pago Pago: American Samoa Science Fair

Puerto Rico
Aguadilla: Regional Mathematics Fair
Cayey: Radians Science and Engineering Fair
Humacao: Humacao Regional Science Fair
Rio Piedras: San Juan Archdiocesan Regional Science Fair

Virgin Islands
St. Croix: St. Croix Science Fair

Canada

Montreal, Quebec Montreal Regional Science Fair
Regina CWSF/Waterloo-Wellington/Lambton
Regina CWSF/Central and Eastern Newfoundland/Montreal
Regina CWSF/Quebec Regional Science Fair
Regina CWSF/Greater Vancouver
Regina CWSF/Toronto/Durham/Timmins
St. Catharines, Ontario Niagara Regional Science and Engineering Fair

APPENDIX B

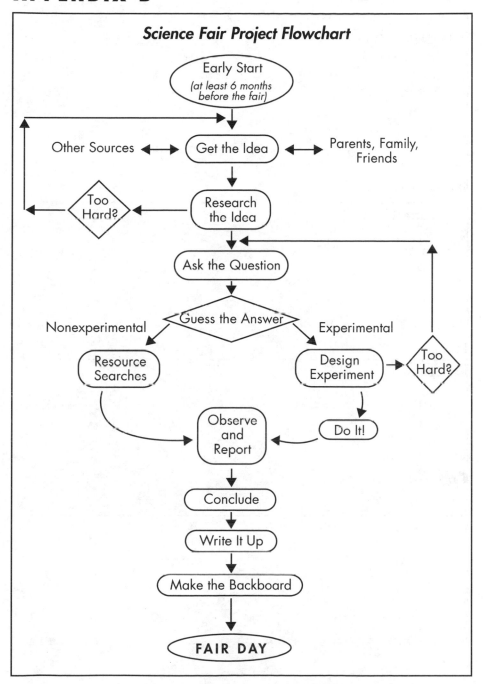

Science Fair Project Flowchart

Early Start
(at least 6 months before the fair)

Other Sources ⟷ Get the Idea ⟷ Parents, Family, Friends

Too Hard? ← Research the Idea

Ask the Question

Guess the Answer

Nonexperimental — Resource Searches

Experimental — Design Experiment → Too Hard?

Do It!

Observe and Report

Conclude

Write It Up

Make the Backboard

FAIR DAY

APPENDIX C

TOPIC:
Source 1
Title:
Author:
Notes:

Source 2
Title:
Author:
Notes:

Source 3
Title:
Author:
Notes:

Source 4
Title:
Author:
Notes:

APPENDIX D

These are some of the sites we have found on the Internet. None of these sites provide an entire project to download. Some of these sites are an aid to doing a project, some give samples, and some are general information. Enjoy!

http://kidscience.miningco.com/mbody.htm

http://www.stemnet.nf.ca/~jbarron/scifair.html

http://www.ipl.org/youth/projectguide/

http://www.sciserv.org/iisef.htm

http://trfn.clpgh.org/Education/K12/homework/scifair.html

http://physics.usc.edu/~gould/ScienceFairs/

http://thegardenhelper.com/schoolwork.html

http://atlas.ksc.nasa.gov/education/general/scifair.html

http://www.parkmaitland.org/sciencefair/index.html

http://www.halcyon.com/sciclub/cgi-pvt/scifair/guestbook.html

http://www.isd77.k12.mn.us/resources/cf/steps.html

http://www.ctcase.org/sciencefairs.html

http://www.gallaudet.edu/~mssdsci/scifairs.html

http://www.gsn.org/cf/index.html

EASTERN NEWFOUNDLAND REGIONAL SCIENCE FAIR

Judging Form for Experimental Projects

TOTAL MARK: _____

Project #: _____ Language: _____
Entrant: _____
Partner: _____
Category: _____ Division: _____ Type: _____
Project Title: _____

Experimental Project — An investigation undertaken to test a specific hypothesis using experiments. Experimental variables, if identified, are controlled.

Part A: Scientific Thought (Maximum 45 marks)

Level 1 - Duplication of a known experiment to confirm the hypothesis. The hypothesis is totally predictable.

5 MARKS MANDATORY (Maximum 15/45)

+ 0 1 2 3 4 5 6 7 8 9 10

Level 2 - Extend a known experiment through modification of procedures, data gathering and application.

15 MARKS MANDATORY (Maximum 25/45)

+ 0 1 2 3 4 5 6 7 8 9 10

Level 3 - Devise and carry out an original experiment with controls. Variables are identified. Some significant variables are controlled. Data analysis includes graphic representation with simple statistics.

25 MARKS MANDATORY (Maximum 35/45)

+ 0 1 2 3 4 5 6 7 8 9 10

Level 4 - Devise and carry out original experimental research which attempts to control or investigate most significant variables. Data analysis includes statistical analysis.

35 MARKS MANDATORY (Maximum 45/45)

+ 0 1 2 3 4 5 6 7 8 9 10

Part B: Original Creativity (Maximum 25 marks)

1. Topic originality	5	4	3	2	1	0
2. Originality in approach	5	4	3	2	1	0
3. Resourceful use of equipment and information services	5	4	3	2	1	0
4. Creativity in interpretation of data	5	4	3	2	1	0
5. Judge's discretion	5	4	3	2	1	0

Part C: Skill (Maximum 10 marks)

1. Necessary scientific skill shown	3	2	1	0
2. Exhibit well constructed	3	2	1	0
3. Material prepared independently	3	2	1	0
4. Judge's discretion		2	1	0

Part D: Dramatic Value (Maximum 10 marks)

1. Layout logical and self-explanatory	3	2	1	0
2. Exhibit attractive	3	2	1	0
3. Presentation by student clear, logical, enthusiastic	3	2	1	0
4. Judge's discretion		2	1	0

Part E: Project Summary (Maximum 10 marks)

1. Has all the required information been provided?	3	2	1	0
2. Is the information in the specified format?	3	2	1	0
3. Is the information presented clearly with continuity?	3	2	1	0
4. Does the summary accurately reflect the actual project?		2	1	0
5. Presentation (neatness, grammar, spelling in report)		2	1	0

APPENDIX F

Science Fair Projects

These are projects taken from the Science Fairs Homepage www.stem-net.nf.ca/~jbarron/scifair.html

They are originally from a publication of the Youth Science Foundation, a list of previously used project ideas (YSF 1980).

Primary Projects—Grades One to Four

A Cow's Horns	Insects: Bad Guys or Good Guys?
Our Friend the Wren	Parts of a Bird
A Model Arm	The Solar System
Care and Feeding of Birds	Parts of a Flower
Butterflies	Parts of the Eye
Cotton Culture	Parts of a Hen's Egg
Growing Trees	Primitive Animals
Ecology of our School Grounds	Parts of a Horse
Human Eye	The Age of Reptiles
Food Values	Plants Grow Toward Light
Inside Fruit	Earthworms
Fat is Fatal	Prehistoric Animals
Leaf Characteristics	Ants
Habitat Studies	Seed Collection
Flies	Miniature Greenhouse
How a Tooth Decays	Rats Are Eating Your Food
Migration of Birds	Germination
How Plants Reproduce	The Beaver
Nutrition	Coal

The Coyote

Natural Gas

The Jack Rabbit

Clouds

The Prairie Chicken

Rain and Snow

Tales, Trails & Tails of the Wolf

Animal Movement

The Earthworm, Master Plowman

The Sun, Moon and Earth

The Human Heart

Rock Collection

The Life Story of a Tree

Food Collection

Carbon Dioxide and Man

Flower Collection (Annuals, Biennials, Perennials)

The Way to Test for Food Starch

A Balanced Aquarium

Nuts, Cones, Wild Flowers, Shells, Insects

Bees

A Garden

A Healthy Breakfast

Primary Projects—Grades Four to Six

A Bell System

History of Shells

A Chemical Change

Types of Fuels

A Crystal Radio Set

Heat Can Produce Electricity

Action of a Solenoid

Fire Must Have Air to Burn

A Door Chime

How Electricity is Made

Air Currents

Climate

An Electronic Map of Canada

Electric Eye

Weaving and Sewing Techniques

ElectroMagnetism

A Projector

Fluorescent Lights

Measuring the Ocean Depths

Functions of a Camera

Measuring Outer Space

Glass and its Uses

Model Airplanes

How to Develop a Picture

Molding

How Traffic Signals Work

Most Liquids Contain either Acid or Alkali

Inside a Cave

Manufacturing Machinery

Operation of a Doorbell

Mercury

Parts of an Electric Motor

Minerals: Origin, Distribution

The Arc Light

A String Pump in Action

Our Community Planning

A Weather Station

Our Solar System

A Cotton Gin

Phases of the Moon

A Wheat Elevator

Printing and its Value to Man

Cross Section of a Volcano

Salt and its Uses

Cross Section of an Oil Well

Weather and Man

Cross Section of the Earth

Simple Machines

Distillation of Water

Sound

Power and Food from the Sea

Sulfur

An ElectroMagnet

Fingerprinting

Expansion and Contracting of Liquids

The Telegraph Key

Man's Natural Resources

Which Metals Conduct Heat?

Rotation of Planets

The Telephone

Sending Messages by Electricity

Train Signal

Light

Water Finds its Own Level

Snowflakes

Water Supports Heavy Weights

Space Travel is Coming

An Electromagnetic Crane

Space Problems in Gravity

Machines and Tools

Steam Propulsion

Parts of a Sailboat

Steam Turbines

Parts of a Windmill

The Fulcrum and the Lever

Polar Constellations

The Planets

Principles of a Transformer

Water is Compound of Hydrogen and Oxygen

Products of Oil

Jet Propulsion, Natural and Man-Made

Working Principles of a Gasoline Engine

The Quartz Family

Working of a Telegraph

A Reed Basket

Workings of a Television

Camera

Workings of an Irrigation Pump

Air Pressure in a Mercury Barometer

Machines Made Work Easier

Astronomy

Birth of a Balloon

The Blinker Light

Causes of the Seasons

Bulbs in Series and Parallel

The Climate of Your Own Home

Chemurgy

Chlorophyll

Canals and Locks

Contour Mapping

Using a Compass

Weather Instruments—forecasting

Minerals

How Accurate are Homemade Weather Instruments?

Water Cycle

Rocks

How Hard are Various Types of Rocks and Minerals?

Mining—coal, iron ore, etc.

Petroleum and Oil

Erosion—what causes erosion? How can soil erosion be prevented?

Air Pollution—causes and cleanup

Water Pollution

Forces Changing the Earth's Surface

Precipitation

Water Filtration

The City of the Future

A Study of a Stream

The River

Topographic Mapping

Tides

Earthquakes and Associated Measuring Devices

Hurricanes

Collect and Identify Minerals and Rocks Exposed in Local Area

Floods

Experimental Projects—Grades Four to Six

Electricity

Demonstration of principles; how is current affected by type of conductor, temperature, filament, etc.

Compare electromagnets for strength, wires for conductivity

Principles of fluorescent lights; how do they compare with filament bulbs, in effectiveness and cost

Physics

How metals compare in conducting heat

How metals compare in density and buoyancy

Efficiency of different types of steam engines

How does the amount of oxygen affect the rate of burning?

Does temperature affect solubility?

Are some substances more soluble than others?

How do airplanes fly? What is the best wing shape?

How do waves carry energy?

How do magnets work? How are they made?

Compare densities of different gases

How light is affected passing through water, e.g., viewing objects underwater, formation of rainbows

What limits the speed of boat, truck transport?

Chemistry

Chemical change and the factors that affect the rate such as heat, light, and catalysts

Acid and basic solutions, how are they produced, how can they be modified; practical considerations in soil, lakes, food; acid or basic solutions around the house

Factors affecting the making of glass

The effects of salts on the freezing point of water and other liquids

Meteorology

Day length—record length of days and nights over a period of time; what effects do the changes have on things like household plants, pets, etc.

Air movement—is air in your house the same temperature at floor level and near the ceiling?

How could you spread heat more evenly throughout the house?

Dew and frost—does it form on clear or cloudy nights? What other factors increase the amount of dew? Can you measure how much dew is formed in a square meter?

Temperature—How does the temperature change during the day? What time is usually the warmest?

Can you construct your own thermometer to keep your own records?

Rain—how does a rain gauge work? Measure the rainfall over a period of time and compare it with the daily weather reports. Principles of cloud seeding and other weather modification

Biology

Insects—personal observations on life cycle, feeding habitats, population, flies, bees, butterflies

Nutrition—plants and fertilizer

Studies—how pet mice respond to different types of food (pellets, crushed, solid)

How do plants get nitrogen

Plants—why do plants grow toward light? The effects of gravity on seed germination. How water moves through the plant. How plants reproduce and factors that affect the process. Why do plants move?

Soil—the importance of earthworms to soil and plants. The effect of soil components and organic matter on growth of plants

Field studies—plant and animal life in the school grounds, a creek or stream, a grassy field, a tree, a home garden, a balanced aquarium, during winter. Diets of various animals

Experimental Projects—Grades Seven to Nine

Physics

Fire and burning—what factors affect burning?

Fuels and their efficiency in producing energy

Musical instruments—the scientific principles behind them

Music vs. Noise—difference

Pendulums—how can a period of a pendulum be increased?

Air pressure–water pressure

Gears—compare efficiencies, effect of different lubricants

Solar furnace

Lenses—effects of curvature, materials on light beams

Can eggs withstand a greater force from one direction than from others?

How strong are nylon fishing lines?

How strong are plastic wraps?

Which homemade airplane design flies best?

What factors affect the bounce of a dropped ball?

How do compression and tension make things strong?

How strong is a toothpick?

Which type of lawn sprinkler works best?

Which type/size of light bulb produces the most light?

How can the strength of light be measured? The effect on degradable materials

Which materials can be charged with static electricity?

Which battery lasts the longest? How can power be increased?

What affects light reflection? Refraction and diffraction of light?

Spectrum and color production—prisms

How is sound produced? What affects the pitch of sound? What affects the volume of sound? How would you measure the velocity of sound?

Electric motors—principles and factors affecting their efficiency

Electric circuits—factors affecting voltage, amperage, resistance

Magnets and electromagnets—What affects the strength of an electromagnet?

Buzzers and bells and alarms

Radios

Internal combustion engines

Heat convection—radiation of heat

Insulation—best materials, thickness

How is paint affected by temperature changes?

Elasticity of rubber; effect of glue

Engineering/Physics

Use of solar energy—design and construct solar cookers, solar panels, etc.

Design a strong bridge, an energy efficient home

Efficient use of renewable energy resources, e.g., wood, wind

Determine the accuracy of various thermometers

How much heat is required to raise the temperature of various substances by an equal amount?

Principle of energy conservation

Comparing active and passive solar energy systems in cost and efficiency

Meteorology

Snow—what happens when it melts; what does it contain; structure of snowflakes; life in a snowbank

Sky color—account for differences in color at different times

Wind and clouds—what are the common wind patterns in your area and why? Is cloud formation related to height, weather systems, and temperature? Study and record how clouds relate to weather patterns

Water levels—study and record varying levels over the year in a body of water; account for differences and the results on the surrounding environment

Dew formation—how much is formed on a square meter for a period of time; account for variations

Wind—does wind travel at the same speeds and in the same directions at different heights?

Frost formation—what must the temperature be to form frost; what are the effects of humidity?

What is the makeup of frost and dew?

Evaporation—which affects the rate of evaporation most—temperature, humidity, wind speed, or other factors?

Rain—can you measure the speed and force of raindrops? What is the effect on soil, with and without ground cover? Could you simulate the effect of rain?

Heat retention—does fresh water hold heat longer than salt water? How does water compare to land and what effect does this have on the weather? What factors affect the cooling of land?

Sunlight—how do different surfaces affect the amount of sunlight reflected and absorbed?

Design a method of measuring how much sunshine is available each day

Humidity—can you collect the amounts of water in the air at different temperatures?

Temperature—what is the difference between direct sun and in the shade? Is the difference constant?

Weather records—design and build an automatic recording weather device. Test it over a period of time

Effects of humidity—what happens to hair during periods of changing humidity? How does human hair compare to that of other animals? How do other materials compare in expansion and contraction?

Chemistry

Effects of temperature on viscosity of oil, chemical reactions, Brownian motion, burning of different materials

Everyday activities that illustrate chemical principles

Chemical reactions that produce energy or that require energy

Testing of consumer products—glues, stain removers, antiseptics, mouthwash, detergents, paper towels, making salt water potable, removal of pollutants

Effects of sunlight on rubber, ink, paper

Effects of increased concentrations on the rate of chemical reactions

Compare the pH levels in mouths of various animals and humans at different times of the day

Compare the surface tension of various liquids

Dealing with chemical spills from industry

Analyzing snow and rain for pollutants; samples from different locations

Effects of temperature on density of gases

Effects of salt and other contaminants on rate of rusting

Growing crystals—factors that affect the rate and the size

Can you obtain water from ink, vinegar, milk?

What effects do different amounts of exercise have on the production of carbon dioxide in humans?

Analyze soil samples for their components, ability to hold moisture, fertility and pH

Does the amount of particle pollution vary with distance from a road, with location, with height? Determine types of particles found in pollution fallout

Catalysts—how they work and why; commercial applicants and problems

Fire extinguishers—principles of operation and factors affecting their efficient use

How do acids react with different metals under varying conditions?

Identify different metals by the color of flame when they burn

Can you devise an experiment that will list metals in order of their activity, from the most potassium to the least active ore gold

Electroplating—the principles, how different metals can be used, and the practical applications

Botany

Germination—how monocots and dicots differ. The effects of heat, light, carbon dioxide, pH level, etc. on germination rate

Photosynthesis—factors affecting the rate of photosynthesis temperature, light intensity, water, carbon dioxide. Part of light spectrum used in photosynthesis

Leaf—Do the numbers and sizes of stomata vary with different plants? What happens if stomata are covered and why?

Roots—how much water is used by different plants? What is the effect of temperature, sunlight, etc., on the use of water (transpiration)? How do different types of soils affect the ability of roots to anchor plants? What factors encourage root growth and what is the effect of water, oxygen, soil type, minerals on root growth?

Plant growth—determine the effects of various nutrients, amounts of water, hours of sunlight, strength of weed killer, temperature, pollutants, pH levels on plant growth and crop yields. Can plants live without oxygen, carbon dioxide? What percentage of various plants is water?

Genetic studies—connections between hair and eye color, sex and left handedness, hair color and strength. Family studies on inheritance

What conditions are favorable for: fungus growth, e.g., yeast, mold, mildew diseases; mushroom production; growing brine shrimp; algae growth; bacteria growth or control; mutations; rooting cuttings; the survival of planaria; the growth of nitrogen fixing bacteria; lichen growth

Field studies—effects of herbicide spraying, acid rain in a lake, auto exhausts on a roadside, SO_2 emissions on plants, under hydro lines. Types of bacteria found around the home, on the body, in soil of different types

Reactions of protozoa to changes in the environment

The preferred pH level in the soil for various plants

Senior Projects—Grades Nine to Twelve

Chemistry

Design considerations for "solar heated" homes

Design considerations for "solar-cell" powered homes

A study of propeller designs for wind generators

Production of electrical energy from mechanical sources

Study of efficient home insulation

Comparing insulative properties of various natural and commercial insulators

The effect of landscaping and architecture on energy consumption

Efficiency studies on transformers

The effect of temperature on resistance

Study of formation of images on a TV tube

Efficiency studies of L.E.D (light emitting diodes)

L.E.D illumination versus incandescent illumination in practice

Voice communication with infrared light and fiber optics

Find the maximum speed in fiber-optic links

Study of various phosphors in fluorescent lighting

Structure versus strength in dams

Testing and comparing consumer products

Earth Science, Meterology

Observations of experimentally induced seismic waves

Is there a relation between sunspot cycles and earthquakes?

Observations of geomorphic factors in the local areas

Tracing glacial till fragments to local rock outcrops

Exploring methods of controlling erosion

Fossil studies in limestone and other rocks

A study of phosphorescence as a tool for geologists

Comparison of the load bearing strength for different soils

Observations of fluctuations in stream flow following rain

Study of air tides: phases of the moon versus barometric pressure

Effects of weather on human emotions

Changes in snow density and other characteristics with time

The factors affecting ice patterns on glass

Study of the relation between wind direction and temperature inversions

A study of small scale wind currents around buildings

Observations of local anomalies in the earth's magnetic field

Environmental Science

The study of flora in a given region

Observations of urban wildlife

Study of adaptations of city flora to smog

An ecological study of the animal and plant populations occupying the same tree

The effects of crowding (with the same or other species) on a certain plant

Annual variations in the ecology of a body of water

A study of a shoreline

Observations of the spread of Dutch Elm disease

A study of the relation between soil type and vegetation

A study of the relation between vegetation and insects

Monitoring the changes in wildlife caused by human encroachment

The study of the impact of pollution on an ecosystem

A study of water pollution from feed lot farms

Tracing chemical (e.g., DDT) concentrations in successive food chain levels

Ozone destruction experiments

A study of air purification methods

Efficient methods of breaking down crude oil in seawater

Experimenting with microbial degradation of petroleum

Experimenting with biodegradability

Finding efficient methods of harvesting and using plankton

Find an ink that would decompose for recycling paper

Computer Science

Studies of storage/retrieval techniques for computer systems

Handling of data transfer between I/O devices

Data manipulation and information management techniques and procedures

Applications in education using the computer as an education tool

Compiler design

Statistics and random number problems

Simulation of nonscience areas, e.g., history, life on other planets

A programmable processing unit design, function and operation

Developing a video game

Pascal programming tools

Developing a program to write a new custom program

Use of computers in managing industrial processes

Using computers to help people do what they want to do

Biology

The effect of sound on plants

Plants in different environments (light intensity, color)

The effect of nicotine, air, yeast on mold growth

Factors affecting the strength of hair, the growth of bacteria, molds, or yeast

Experiment with hydroponics

Use seedlings started from seed with three types of soil and different rates of fertilizer

The effectiveness of antiseptics and soaps on household bacteria

The effect of air pollution on algae, protozoa, fish, insects, or mosses and lichens

Comparing types of artificial light on plant growth

Conditions necessary for the life of brine shrimp

The commercial uses of algae methods of production

Producing mutations in bacteria, yeast, protozoa, or molds

Best conditions for mushroom production, growth of ferns

The effects of ultrasonic antibiotics, temperature changes on bacteria count

Microbial antagonism

Reaction of paramecia, planaria, to pH, light, and temperature conditions

Plant tropisms and growth hormones

Transpiration rates for different plants and conditions

Sugar level in plant sap at different times and dates

Using radioisotopes to study uptake of plant nutrients

A study of territoriality in mice

A study of the cleaning habits of mice

Observation of conditioned responses in different animals

A study of animal phosphorescence and other bioluminescences

Learning and perception in animals and humans

Studies of memory span and memory retention

Age versus learning ability

A study of the relation between physical exercise and learning ability

Is audio or visual information better remembered

The effect of bleaching and dyeing on hair

A study of the percentage of DNA (by weight) in different species

Factors affecting the enzyme's reaction rates

Genetic variations across a sansevieria leaf

Factors affecting seed germination (e.g., soil temperature, pH)

Root formation in cuttings versus lighting conditions

Factors affecting flowering

Study of sterility in plant hybrids (F1 and F2)

Comparison of different plant's ability to add humus to the soil

Factors affecting nodule formation in legumes

Can household compounds (e.g., tea) be used to promote good health in plants

Effects of cigarette smoke on the growth of plants

The effects of water impurities on plant growth

The effects of phosphates on aquatic plants

Effect of mineral deficiencies on protein content in soybeans

The effect of excess salinity on plants

A study of the tumors produced in plants by *Agrobacterium tumifacieus*

The effect of polarized light on plant growth direction

The effects of solar activity on plant growth

Tracing solar activity cycles in tree growth rings

The effects of electric fields on plants

The effects of magnetic fields on plant growth

Effects of magnetism on the size and frequency of blooms and fruits

Does magnetizing seeds before planting affect growth

The effects of X ray and other radiation on plants

The effect of music of varying types and duration on plants

Organic fertilizer versus chemical fertilizer

Study of population fluctuations in insects

A study of toxicity of insecticides versus temperature

Is polarized light the guidance system for foraging ants?

A study of stimuli that attract mosquitos

The factors affecting the rate at which a cricket chirps

Study of insect of animal behavior versus population density

Study of diffusion through cell membranes

Growth of plant and animal cells by cloning

Regeneration in sponges, paramecia, planaria, etc.

Manipulation of vegetative reproduction in plants

Search for near vacuum environment tolerant plants

Physical Science

Study of accuracy of calculators

The mathematics of snowflakes

Observational orbit determination of comets, meteors, or other minor planets

The effect of solar activity on radio propagation

Observations of sunspots, flares, and prominences

A study of solar flares through the sudden enhancement of atmospherics

The identification of elements in the solar and stellar spectra

Experimental exploration of the photoelectric effect

Experimenting with electron diffraction

Observations of magnetic permeability of different materials

Comparing magnetic pysteresis for different material

A study of radiation patterns from different antenna types

Factors affecting scent propagation

Factors affecting sound propagation

Factors affecting sound dampening

Index of refraction of liquids versus temperature

Index of refraction of liquids versus amount of additive

A study of infrared qualities of certain solutions

Crystal growth rates versus solution strengths, temperature, etc.

Observation of freezing rates of water for different starting temperatures

Reproduce the Stanley Miller experiments "The Origin of Life"

Find the optimal gas mixture for a Stanley Miller experiment

Experimenting with various separation techniques (e.g., electrophoresis)

A study of catalyzed reactions

A study of saponification reactions

A study of esterification reactions

The physics of ski waves

BIBLIOGRAPHY

Asimov, I. and A.D. Fredericks. 1990. *The complete science fair handbook*. Glenview, Ill.: Good Year Books, Inc.

Barron, J. 1997. *An investigation into the sources for ideas and research of students participating at the regional science fair level*. Masters Thesis, Memorial University Press.

Bellipanni, L.J., D.R. Cotten, and J.M. Kirkwood. 1984. In the balance. *Science and Children*. 21(5). Reproduced in Watt, S.L., ed. 1988. *Science Fairs and Projects K–8*. Washington, D.C.: National Science Teachers Association, 36.

Bombaugh, R. 1987. Mastering the science fair. *Science Scope*. 11(2). Reproduced in Watt, S.L., ed. 1988. *Science Fairs and Projects K–8*. Washington, D.C.: National Science Teachers Association, 18–20.

Bosak, S.V. 1991. *Science Is* . . . 2d ed. Richmond Hill, Ontario: Scholastic Canada Limited.

Chiappetta, E.L. and B.K. Foots. 1984. Does the science fair do what it should? *The Science Teacher.* 51(8). Reproduced in Watt, S.L., ed. 1988. *Science Fairs and Projects K–8.* Washington, D.C.: National Science Teachers Association.

Collins, B.K. 1981. Independent projects—an organized approach. *The American Biology Teacher.* 43(8). November.

Connors, W., et al. 1988. *Science fare.* Markham Ontario: Sciencefare Press, Inc.

Daab, M.J. 1988. *Improving fifth grade students' participation in and attitudes toward the science fair through guided instruction.* Edited by D. Practicum. Fort Lauderdale, Fla.: Nova University.

DeBruin, J., et al. 1993. Science investigations mentorship program. *Science and Children.* 30(6): 20–22.

Fields, S. 1987. Introducing science research to elementary school children. *Science and Children.* 24(1): 18–20.

Foster, G.C. 1983. Oh no! A science project. *Science and Children.* 21(3): 20–22.

Giese, R.N., J.H. Cothron, and R.J. Rezba. 1992. Take the search out of research. *The Science Teacher.* 59(1), January.

Hamrick, L. and H. Harty. 1983. Science fairs: a primer for parents. *Science and Children.* 20(5): 23–25.

Hansen, B. 1983. Planning a fair with flair. *Science and Children.* 20(5). Reproduced in Watt, S.L., ed. 1988. *Science Fairs and Projects K–8.* Washington, D.C.: National Science Teachers Association, 14–16.

Henderson, S.A. 1983. Did Billy Gene do this project himself? *Science and Children.* 20(5). Reproduced in Watt, S.L., ed. 1988. *Science Fairs and Projects K–8.* Washington, D.C.: National Science Teachers Association, 48.

Hudson, T. 1994. Developing pupils skills. In Levinson, R. ed. *Teaching Science*. London and New York: Routledge Press.

Knapp, J. 1975. Science fairs in the eighth, seventh, or sixth grades? *Science and Children*. 12(8): 9.

Liebermann, J. Jr. 1988. Using kinetic experiments from the journal of chemical education as the basis for high school science projects. *Journal of Chemical Education*. 65(12), December.

McBurney, W.F. 1978. The science fair: A critique and some suggestions. *American Biology Teacher*. 40(7): 419–422.

McNay, M. 1985. The need to explore: Non-experimental science fair projects. *Science and Children*. 23(2), October.

Mann, J.Z. 1984. *Science Day Guide*. Columbus, Ohio: Ohio Academy of Science.

Nelson, D.J. 1986. Consumer fair: A cure for the science fair blues. *Science Scope*. 10(6). Reproduced in Watt, S.L., ed. 1988. *Science Fairs and Projects K–8*. Washington, D.C.: National Science Teachers Association, 56.

Ohio Academy of Science. 1987. *Why? Student Research*. Columbus, Ohio.

Oswald, R. and C. Preyra. 1990. *Gage Science 8 Making Connections Teacher's Guide*. Toronto, Ontario: Gage Educational Publishing Company.

Pike, L. 1993. Science day at our school. *NTA Science Council Bulletin*. 3(1): 8.

Pryor, C. and A. Pugh. 1987. Science fairs: A family affair. *Science Scope*. 11(2). Reproduced in Watt, S.L., ed. 1988. *Science Fairs and Projects K–8*. Washington, D.C.: National Science Teachers Association, 49.

Pushkin, D. 1987. Science fair projects: Some guidelines for better science. *Journal of Chemical Education*. 64(11): 962–963.

Rao, C.S. 1985. *Science teacher's handbook.* Reprint r-50. Washington D.C.: Information Collection and Exchange Div. Peace Corps.

Regional science fair guidelines. 1980. Ottawa, Canada: YSF.

Riggins, P.C., ed. 1985. *Science fair: It's a blast! A Guide for Junior High Students. Teacher's Guide.* Springfield: Illinois State Board of Education.

Rivard L. 1989. A teacher's guide to science fairing. *School Science and Mathematics.* 89(3): 201–207.

Romjue M.K. and J.J. Clementson. 1992. An alternative science fair. *Science and Children.* 30(2): 22–24.

Scarnati, J.T., et al. 1992. Science Showcase Night. *Science Scope.* 15(7): 38–41.

Science Service. 1998. Intel International Science and Engineering Fair Homepage. www.sciserv.org.

Silver, D. 1994. Science fairs: Tired of the same old, same old? *Science Scope.* 17(5): 12–16.

Simms, W. 1993. Science day for the whole primary and elementary school. *NTA Science Council Bulletin* 3(1): 7.

Sittig, L.H. 1985. Whoever invented the science fair . . . *Science and Children.* 22(5). Reproduced in Watt, S.L., ed. 1988. *Science Fairs and Projects K–8.* Washington, D.C.: National Science Teachers Association, 50.

Stedman, C.H. 1975. Science fairs, model building, and non-science. *Science and Children.* 12(5): 20–22.

GLOSSARY

apparatus: *see* materials.

backboard: A display area used to demonstrate a student's project. It is a freestanding board, usually divided into three sections. Rules govern the materials used for backboards so as to reduce fire danger.

baseline result: The results obtained by running the experiment without changing any of the variables. This result provides a baseline against which to measure the results of the experiment.

brainstorming: A process that involves writing down the main idea and then writing all the things that come to mind from the main idea. In science we use brainstorming to determine everything we do and don't know about an idea.

carcinogens: Materials known to cause cancers. *See also* controlled substances.

consumer project: A test of commercially available products, where the effects of different brands are compared. Some debate whether this sort of project is truly experimental, but it is allowed at many science fairs.

control: An experiment run without changing any of the variables. A control yields a baseline result.

controlled substances: Substances such as drugs, alcohol, or tobacco, or items such as weapons or explosives, the use or possession of which is governed by state or federal law. *See also* hazardous materials.

controlling variables: Making sure we don't change any other parts of the experiment.

designated supervisor: A person, not necessarily a scientist, who is trained in the use of hazardous materials. A designated supervisor is required for experiments involving hazardous materials. *See also* qualified scientist.

discrepant event: A result that was not expected.

DNA: Deoxyribonucleic acid. The basic building blocks of life.

enzyme: A protein that helps speed up the natural breakdown of various substances.

experimental project: A project involving an experiment where something is tested to achieve a predicted result.

hazardous materials: Materials, such as explosives, radioactive materials, and firearms; also hazardous chemicals such as pesticides, carcinogens, and toxic or flammable chemicals.

ISEF: International Science and Engineering Fair. The biggest and most prestigious international science fair. If a regional fair is affiliated with the ISEF, winners in that fair can compete in the ISEF. The ISEF also sets useful guidelines that even unaffiliated fairs often follow.

materials: The items needed to carry out an experiment. Also called apparatus.

nonexperimental project: A project that involves only research into a subject. *See also* experimental project.

paraphrasing: Taking something you have read and putting it into your own words.

pathogen/pathogenic agent: Something capable of causing disease.

placebo: A fake pill that is given to patients during tests of new medicines. Often used as a control.

plagiarism: Taking another's work and calling it one's own.

qualified scientist: A trained scientist who helps science fair participants. Some science fair projects, such as those involving human subjects, vertebrate animals, pathogens, controlled substances, recombinant DNA, or human/animal tissue, require the help of a qualified scientist. *See also* designated supervisor.

quantify: To give a numeric value to an observation.

rDNA: recombinant deoxyribonucleic acid. A form of DNA with some strands removed and combined with strands from other DNA.

science fair: An occasion for the display and evaluation of student research projects.

science vs. technology: Science is mostly theory and technology is mostly the application of that theory.

scientific review committee: A group of people who are responsible for ensuring that a science fair project is within the rules (and any applicable laws). Can also be called a safety committee, an ethical research committee, or a project review committee

secondary sourcing: Finding other research sources by looking at the bibliography or reference list of an original source.

S.I.: Système Internationale (International System in English).

size limitation: The total size a project can occupy is 30 inches deep, 48 inches wide, and 108 inches high including the table it is resting on.

technology: *See* science vs. technology.

variables: The things in the experiment that we change so we can observe what happens.

vertebrate animals: Animals that have a backbone.

INDEX